Advances in Intelligent and Soft Computing 81

Editor-in-Chief: J. Kacprzyk

AF816223

Advances in Intelligent and Soft Computing

Editor-in-Chief

Prof. Janusz Kacprzyk
Systems Research Institute
Polish Academy of Sciences
ul. Newelska 6
01-447 Warsaw
Poland
E-mail: kacprzyk@ibspan.waw.pl

Further volumes of this series can be found on our homepage: springer.com

Vol. 67. V. Snášel, P.S. Szczepaniak, A. Abraham, J. Kacprzyk (Eds.)
Advances in Intelligent Web Mastering - 2, 2010
ISBN 978-3-642-10686-6

Vol. 68. V.-N. Huynh, Y. Nakamori, J. Lawry, M. Inuiguchi (Eds.)
Integrated Uncertainty Management and Applications, 2010
ISBN 978-3-642-11959-0

Vol. 69. E. Piętka and J. Kawa (Eds.)
Information Technologies in Biomedicine, 2010
ISBN 978-3-642-13104-2

Vol. 70. Y. Demazeau, F. Dignum, J.M. Corchado, J. Bajo Pérez (Eds.)
Advances in Practical Applications of Agents and Multiagent Systems, 2010
ISBN 978-3-642-12383-2

Vol. 71. Y. Demazeau, F. Dignum, J.M. Corchado, J. Bajo, R. Corchuelo, E. Corchado, F. Fernández-Riverola, V.J. Julián, P. Pawlewski, A. Campbell (Eds.)
Trends in Practical Applications of Agents and Multiagent Systems, 2010
ISBN 978-3-642-12432-7

Vol. 72. J.C. Augusto, J.M. Corchado, P. Novais, C. Analide (Eds.)
Ambient Intelligence and Future Trends, 2010
ISBN 978-3-642-13267-4

Vol. 73. J.M. Corchado, P. Novais, C. Analide, J. Sedano (Eds.)
Soft Computing Models in Industrial and Environmental Applications, 5th International Workshop (SOCO 2010), 2010
ISBN 978-3-642-13160-8

Vol. 74. M.P. Rocha, F.F. Riverola, H. Shatkay, J.M. Corchado (Eds.)
Advances in Bioinformatics, 2010
ISBN 978-3-642-13213-1

Vol. 75. X.Z. Gao, A. Gaspar-Cunha, M. Köppen, G. Schaefer, and J. Wang (Eds.)
Soft Computing in Industrial Applications, 2010
ISBN 978-3-642-11281-2

Vol. 76. T. Bastiaens, U. Baumöl, and B.J. Krämer (Eds.)
On Collective Intelligence, 2010
ISBN 978-3-642-14480-6

Vol. 77. C. Borgelt, G. González-Rodríguez, W. Trutschnig, M.A. Lubiano, M.Á. Gil, P. Grzegorzewski, and O. Hryniewicz (Eds.)
Combining Soft Computing and Statistical Methods in Data Analysis, 2010
ISBN 978-3-642-14745-6

Vol. 78. B.-Y. Cao, G.-J. Wang, S.-Z. Guo, and S.-L. Chen (Eds.)
Fuzzy Information and Engineering 2010
ISBN 978-3-642-14879-8

Vol. 79. A.P. de Leon F. de Carvalho, S. Rodríguez-González, J.F. De Paz Santana, and J.M. Corchado Rodríguez (Eds.)
Distributed Computing and Artificial Intelligence, 2010
ISBN 978-3-642-14882-8

Vol. 80. N.T. Nguyen, A. Zgrzywa, and A. Czyzewski (Eds.)
Advances in Multimedia and Network Information System Technologies, 2010
ISBN 978-3-642-14988-7

Vol. 81. J. Düh, H. Hufnagl, E. Juritsch, R. Pfliegl, H.-K. Schimany, and Hans Schönegger (Eds.)
Data and Mobility, 2010
ISBN 978-3-642-15502-4

Julia Düh, Hartwig Hufnagl,
Erhard Juritsch, Reinhard Pfliegl,
Helmut-Klaus Schimany,
and Hans Schönegger (Eds.)

Data and Mobility

Transforming Information into Intelligent Traffic and Transportation Services
Proceedings of the Lakeside Conference 2010

Springer

Editors

Dipl.Ing. Julia Düh
p.A. Austria Tech
Donau-City-StraSSe 1
1220 Wien
Austria
E-mail: julia.dueh@austriatech.org

Mag. Hartwig Hufnagl
p.A. ATTC
Zieglergasse 6/DG
1070 Wien
Austria
E-mail: hartwig.hufnagl@asfinag.at

Dr. Erhard Juritsch
p.A. Lakeside Science &
Technology Park GmbH
Lakeside B01
9020 Klagenfurt
Austria
E-mail: juritsch@kwf.at

Dipl.Ing. Dr. Reinhard Pfliegl
p.A. Austria Tech
Donau-City-StraSSe 1
1220 Wien
Austria
E-mail: reinhard.pfliegl@austriatech.org

Helmut-Klaus Schimany
MAS MSc
p.A. ATTC
Zieglergasse 6/DG
1070 Wien
Austria
E-mail: helmut-klaus.schimany@oebb.at

Mag. Hans Schönegger
p.A. Lakeside Science &
Technology Park GmbH
Lakeside B01
9020 Klagenfurt
Austria
E-mail: schoenegger@kwf.at

ISBN 978-3-642-15502-4 e-ISBN 978-3-642-15503-1

DOI 10.1007/978-3-642-15503-1

Advances in Intelligent and Soft Computing ISSN 1867-5662

Library of Congress Control Number: 2010933527

© 2010 Springer-Verlag Berlin Heidelberg

This work is subject to copyright. All rights are reserved, whether the whole or part of the material is concerned, specifically the rights of translation, reprinting, reuse of illustrations, recitation, broadcasting, reproduction on microfilm or in any other way, and storage in data banks. Duplication of this publication or parts thereof is permitted only under the provisions of the German Copyright Law of September 9, 1965, in its current version, and permission for use must always be obtained from Springer. Violations are liable for prosecution under the German Copyright Law.

The use of general descriptive names, registered names, trademarks, etc. in this publication does not imply, even in the absence of a specific statement, that such names are exempt from the relevant protective laws and regulations and therefore free for general use.

Typeset & Cover Design: Scientific Publishing Services Pvt. Ltd., Chennai, India.

Printed on acid-free paper
5 4 3 2 1 0
springer.com

Preface

Over the last few years, the local value of mobility and information in our society has grown tremendously. As the importance of Information and Communication Technologies (ICT) increases, we expect more changes in future mobility behavior. This includes not only mobility behavior for the single user, but also for the transportation of goods and infrastructure operators. It will also affect the regulation of resources and political decision-making. Both, data and mobility become more connected. To cope effectively with the anticipated changes, we must expand our focus and take current developments in both areas into account.

The topic of the Lakeside Conference 2010, Data and Mobility – Transforming Information into Intelligent Traffic and Transportation Services, was chosen to underline the importance of information and mobility in transport and to offer an opportunity to discuss and question current activities in this sector. We will consider intermodal concepts and deployments in particular, where data transfer plays a major role, as this could help to reduce the current lack of infrastructure capacity (especially on roads and at airports and seaports). Using modern technologies, traffic management could become more sustainable and efficient.

The Lakeside Conference is, again, organized by a consortium composed of the Lakeside Technology Park, the Austrian Transport Telematic Cluster, AustriaTech and the American Embassy in Austria.

We are proud to announce that certain presentations in this publication by Springer have been chosen because of their excellence and importance in the field of both data and mobility. When deploying technology in the field of transport and traffic, various technological approaches are applied. Only harmonization and the creation of interoperable interfaces can guarantee to exploit the full potential of all of these approaches. The selected papers, ranging from research and economy to industry, give an overview of the latest trends and state-of-the-art developments.

In addition to this, the Lakeside Conference 2010 presents a variety of papers and speeches by international experts. The objective is to show the supporting possibilities of technology in enabling efficient and economically advantageous traffic management of both people and goods. It is of great importance that we consider the amalgamation of the thoughts and approaches of research and development with those of industry, operators and the economy. From our point of view, data and information will become crucial to implementing intermodal, cross-border, transport management in the future.

The editor would like to thank all those involved, the authors, organizers and sponsors for their enormous support, which was provided despite their existing heavy workloads. Their tireless efforts have made the conference an international event of high standing and quality.

On behalf of the editors
Reinhard Pfliegl

Preface

Contents

A Multimedia-Centric Quality Assurance System for Traffic Messages

Roland Tusch, Armin Fuchs, Horst Gutmann, Marian Kogler, Julius Köpke, Laszlo Böszörmenyi, Manfred Harrer, and Thomas Mariacher

Abstract. We present here a technical system for multimedia-centric verification of traffic situations. Traffic messages from various message sources are automatically linked to the video streams of corresponding surveillance cameras in order to support traffic editors in improving the quality of published traffic messages. Video streams can be recorded for certain types of traffic situations and can also be provided as live streams during the verification stage. Both on-demand and live provision of the streams is achieved based on a distributed video scene-recording system that interacts with the video subsection of the Austrian national operator of the highways and motorways (ASFINAG). The situation verification is enabled by a web application which provides (1) an integrated view of the details of a traffic message, (2) a web map illustrating the location and spatial dissemination of the traffic situation, and (3) the recordings and live streams from selected cameras provided by the scene-recording system. Based on the verification result, the traffic editor may then publish traffic situation scenes with enhanced descriptive data about the traffic situation. The system was deployed in the operational environment of ASFINAG and tested by traffic editors at the national radio channel Hitradio Ö3. Evaluation results have shown that this system considerably improves the verification process of the previous system both in terms of speed and accuracy.

Roland Tusch · Armin Fuchs · Horst Gutmann · Marian Kogler ·
Julius Köpke · Laszlo Böszörmenyi
University of Klagenfurt, Institute of Information Technology,
9020 Klagenfurt, Austria
e-mail: {roland.tusch,marian.kogler,julius.koepke,
laszlo.boeszoermenyi}@uni-klu.ac.at, armin.fuchs@aon.at,
horst@zerokspot.com

Manfred Harrer · Thomas Mariacher
ASFINAG Maut Service GmbH, 5020 Salzburg, Austria
e-mail: {manfred.harrer,thomas.mariacher}@asfinag.at

J. Düh et al. (Eds.): Data and Mobility, AISC 81, pp. 1–13, 2010.
springerlink.com © Springer-Verlag Berlin Heidelberg 2010

1 Introduction

Traffic message quality verification is an important issue for operators of road networks. Many client applications for the provided traffic message services and, in particular the road users, often depend heavily on messages requiring a high level of accuracy in terms of the type, time, and location of the described traffic situations [16]. Delivering relevant traffic information quickly to road users has been an active area of research for many years, but ensuring a high level of accuracy for the published traffic information often still requires the involvement of traffic operators or editors in the messaging process.

This paper addresses the issue of improving the quality of traffic messages by assisting traffic editors by providing a multimedia-centric system for traffic message verification and publication. It combines data from existing traffic message sources with a visualization of their location and spatial expansion. It also visualizes the real events on the roads in terms of recorded or live streams from related surveillance cameras. This system is called the *LOOK Platform*, named after an applied research project entitled *LOOK*, which was carried out along with our partner ASFINAG – Austria's national operator of the highways and motorways. It shows the process of traffic message quality verification in an editor-centric, technical and practical manner, as opposed to using metric-based traffic message quality measurements for individual traffic participants, like the approach adopted by the QFCD model presented in [17]. The presented system supports a centralistic verification approach, where a small group of traffic editors verifies existing traffic messages and publishes new ones. Hence, it targets infrastructure-based systems with a large number of installed sensors and automated traffic message sources. This is a complementary approach to the current trends for traffic situation verification in decentralized systems based on inter-vehicle communication, like the Self-Organizing Traffic Information System (SOTIS) [18, 19].

The remainder of this paper is organized as follows. Section 2 describes the architecture of the LOOK Platform, including all its services, applications, and interfaces with the relevant subsections of ASFINAG. Section 3 presents the main results of our initial system evaluation in an operational environment. And finally, the conclusions and future work plans are presented in section 4.

2 System Architecture

Fig. 1. illustrates the architecture of the LOOK Platform, its interfaces to the sensor and video subsections of the Austrian national operator of the highways and motorways, as well as the data flow along the traffic message, sensor and video data processing chain. It consists of seven services and applications operating in four areas of data processing. Each service/application is described briefly below.

2.1 Traffic Message Filter

The Traffic Message Filter (TMF) is the Platform's main mechanism for processing traffic messages. It has an interface with the national highway operator's sensor subsection to periodically import traffic messages, information about road conditions, and weather forecasts already published. The road conditions and weather forecasts, however, are not relevant for this contribution, and hence omitted here. During start-up, this service also imports quasi-static (not frequently updated) data about all of the currently managed highways and motorways, as well as all surveillance cameras available to this system.

All of the processing steps for this service are carried out by the Traffic Message Processor, which provides additional flexible facilities for post-processing the traffic messages. The most powerful of these facilities is the concept of dynamically loadable data sinks. Each system service or application which is interested in consuming the traffic messages processed by the TMF has to implement a predefined *DataSink* interface, and provide the implementation to the TMF. As can be seen in Fig. 1, this is provided for the Scene Recorder (RecordingControllerSink) and Scene Database (ScenePISink) services, since these two services are the next to process the traffic messages.

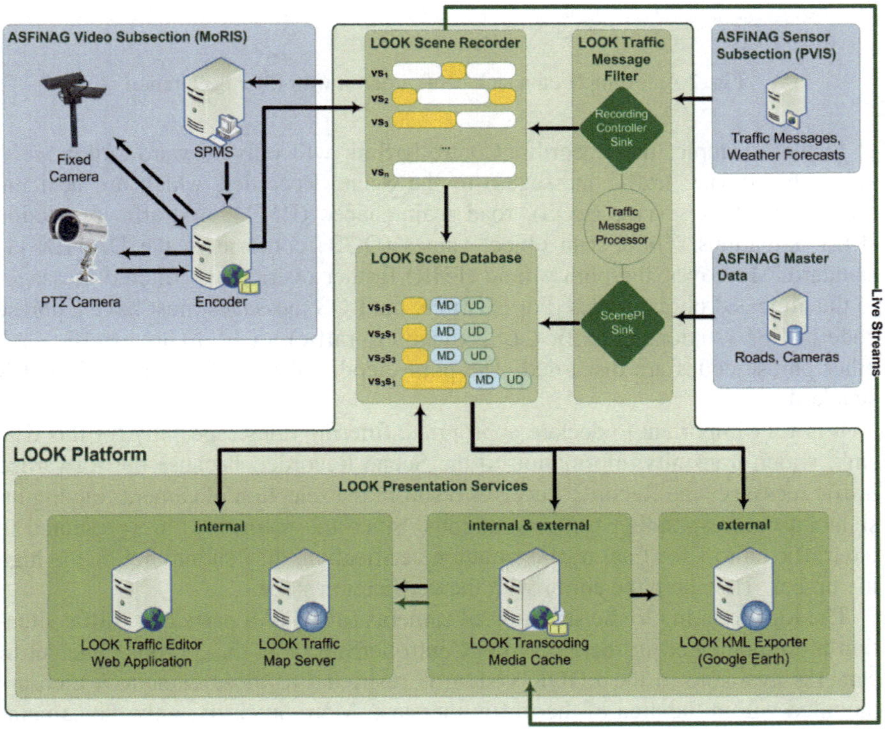

Fig. 1 System overview of the LOOK Platform

After the implementation and provision of a data sink, it has to be linked into the configuration of the TMF, where the new data sink has to be declared. Fig. 2 illustrates such a declaration based on the example of the data sink for the Scene Recorder. This example is of particular interest, since the data sink declaration is also the place where the process of traffic message filtering is controlled.

```
<datasink id="2">
  <name>RecordingControllerSink</name>
  <desc>Posts selected traffic messages to the Recording Controller.</desc>
  <sinkclass>sci.uni_klu.look.tmfilter.sinks.RecordingControllerSink</sinkclass>
  <baseurl>http://recording_controller_host:port/</baseurl>
  <filter>
    <objects>
      <object type="situationelement">
        <restriction>
          <field>DOB</field>
          <values>ACC,LOS,RMT,RES</values>
        </restriction>
        <restriction>
          <field>PHR</field>
          <values>ACI,LS1,LS2,LS3,RCW,CTR,RXB</values>
        </restriction>
      </object>
    </objects>
  </filter>
</datasink>
```

Fig. 2 An example data sink configuration with filter declaration

In this example, the RecordingControllerSink will only forward *situationelement* objects (i.e. traffic messages) to the Scene Recorder, which are accident (ACC), level of service (LOS), road maintenance (RMT), or traffic restriction (RES) situations. These data object values (DOB) conform to the DATEX [1] standard. Moreover, the phrase field (PHR) further restricts the filtered messages to the allocated phrase codes. For example, the LOS messages must have a phrase code of LS1 (stationary traffic), LS2 (queuing traffic), or LS3 (slow traffic). All other phrase codes are discarded. The phrase codes also conform to the DATEX standard.

It is very important to declare appropriate filtering rules especially for this data sink, which transmits information to the Scene Recorder, because the forwarded traffic messages the Recorder receives result in the selection of cameras having insight into corresponding traffic situations. Since the recordings are presented to the traffic editors for final traffic situation verification, they should not be too high in number. They help the editor with the verification stage.

The logic relating to the selection of cameras to record a particular traffic situation is implemented in the RecordingControllerSink. For this purpose, the set of cameras available to the LOOK system is restricted to those cameras which are mounted within the area of the traffic message. More precisely, only the cameras mounted in the traffic message direction and between the road directions are selected. Finally, each filtered traffic message is forwarded to the Scene Recorder together with the set of cameras to record. In the current implementation, the RecordingControllerSink is also responsible for terminating each recording. Hence, the recording time is also configurable for each different type of situation.

2.2 Scene Recorder

The Scene Recorder represents the LOOK interface with the ASFINAG video subsection and is responsible for making recordings about traffic situations, as well as providing camera live streams for the traffic editor. It is the main source of multimedia information in this system, and is itself a distributed system consisting of one *Recording Controller*, and a configurable set of *Recorders*. Fig. 3 illustrates the architecture of the Scene Recorder, including its interfaces and protocols used to communicate with the ASFINAG video subsection and with the other involved services of the LOOK system.

There are two use cases for the Scene Recorder. The first use case is the recordings taken from a given set of cameras for a filtered traffic message by the RecordingControllerSink of the Traffic Message Filter (TMF). The second use case is the provision of a live stream from a given camera for the Transcoding Media Cache (TMC, see section 2.4).

In both use cases, video stream provision is requested by contacting the *Recording Controller*. Depending on the current load and storage capacities of the Recording Subsystem, the Controller selects one available Recorder for taking over the stream provision task. This is accomplished by issuing an RTSP RECORD [2] command to the selected recorder, with the target camera ID(s) received from the TMF or TMC.

In the Recording Subsystem, each Recorder is operated on a separate server to exploit system resources in the best possible manner, and is capable of recording several camera streams and providing one or more live streams at one time. Fig. 4 illustrates the Recorder components and message paths involved in the stream provision process. In the first stage, RTSP commands from the Recording

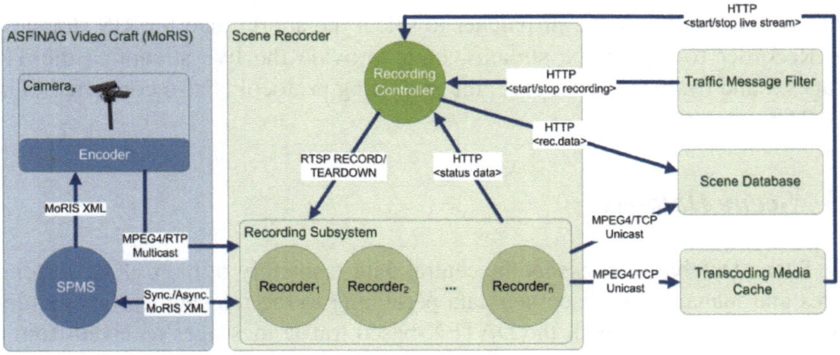

Fig. 3 Architecture and interfaces of the Scene Recorder

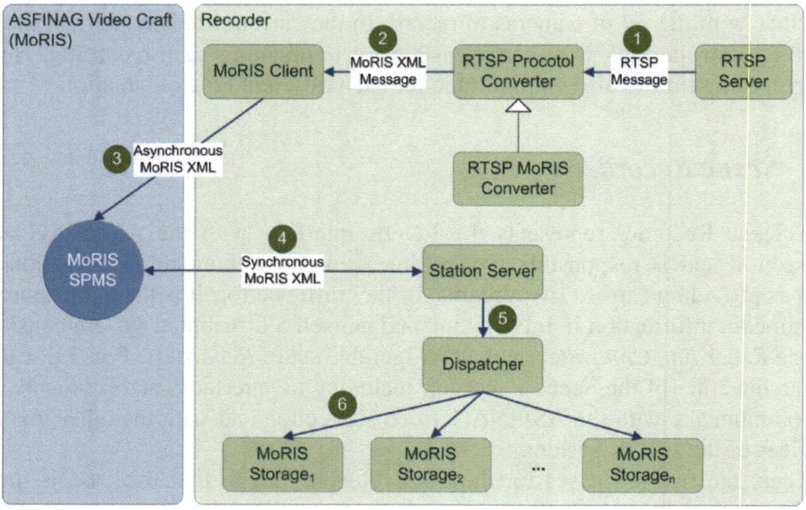

Fig. 4 Recorder components involved in the stream provision process

Controller are converted to MoRIS XML [3] using a protocol converter from RTSP to MoRIS. The conversion process here is 'generic' in the sense that instead of MoRIS XML, any other control and communication protocol (such as the NTCIP protocol family [4]) can be used for systems, which are not based on MoRIS. In the second stage, the converted MoRIS XML message is provided to the Recorder's MoRIS client, which in turn issues the MoRIS command asynchronously to the MoRIS *System Policy Management Server* (SPMS) (stage 3).

In the fourth stage, the SPMS first directs the message to the encoders of the target cameras (not shown in Fig. 4, but in Fig. 3), which in turn provide MPEG4 video elementary streams [5] using the RTP protocol [6, 7] based on certain multicast addresses (where the Recording Controller issued an RTSP RECORD message). Having accomplished this internal MoRIS stage, the SPMS communicates the received multicast addresses to the Station Server. It is the Station Server which performs the final inquiry for the stream provision by joining the multicast addresses and utilizing the dispatcher to select the appropriate MoRIS storages of the Recorder to record the streams, or to provide the live stream to the TMC (stages 5 and 6). In the latter case, the streaming protocol is also converted to unicast TCP.

2.3 *Scene Database*

The Scene Database represents the central data repository of the LOOK system. It stores and manages all processed data pertaining to the relevant traffic situations. These include details about the DATEX-based traffic messages received from the Traffic Message Filter, the scene-recordings provided by the Scene Recorder related to these messages, the additional metadata (tags) provided by the traffic editors via the Traffic Editor Application (see section 2.6), and finally the new publications issued by the traffic editors. Fig. 5 shows the core relations of the Scene Database for managing these data. The central relation is the *scene* relation, which stores the recordings from the Scene Recorder, and relates each recording with the associated traffic message (relation *situationelement*), the camera from which the recording was taken (relation *camera*), the tags that were added by the traffic editors (relation *tag*), and finally the publications for certain users and groups for which the scene has been published by the traffic editors.

The management of the data in the Scene Database is accomplished on three layers. The bottom layer is the logical database layer running a PostgreSQL database instance combined with a file system for the actual video data. There is no direct access for the other services (like the Traffic Message Filter) to these data stores. This access is enabled by an intermediate layer called *Scene Persistence Interface* (ScenePI), which provides abstraction interfaces for all the other services and applications of the system. This is also the reason why the data sink of the TMF for the Scene Database is called *ScenePISink*. It is HTTP-based, utilizes Hibernate 3 [8] for the interaction with the underlying data store, and provides a basic access control for other services. In the case of an admitted query, all of the result data (except video data) is transformed into *JavaScript Object Notation* (JSON) [9] object descriptions instead of XML documents for performance reasons. Finally, the top layer consists of the main presentation and management components for visual traffic situation verification, implemented as the web-based Traffic Editor Application (see section 2.6).

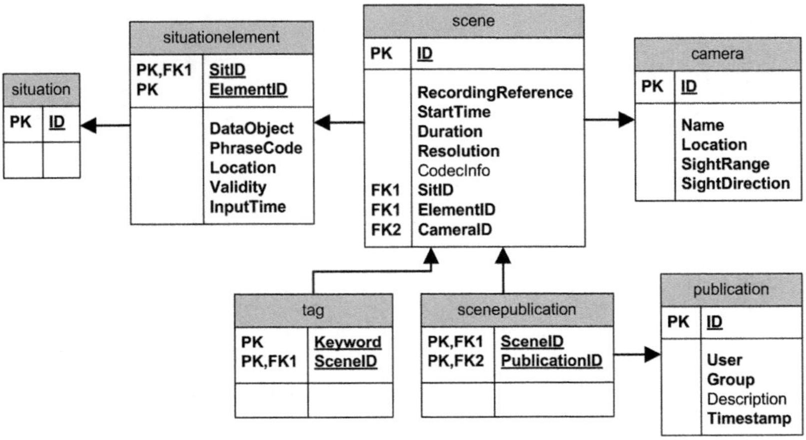

Fig. 5 Core relations of the Scene Database

2.4 *Transcoding Media Cache*

The Transcoding Media Cache (TMC) is an HTTP-based stream transcoding service for converting live camera streams, as well as stream recordings provided by the Scene Recorder from the MPEG4 video elementary stream input format to various output formats. Its primary clients are the web-based Traffic Editor Application (see section 2.6), and the KML Exporter (see section 2.7) for the scenes published by the traffic editors. Fig. 6 depicts the usage scenarios for the LOOK system's central video presentation service. While camera live streams are directly captured as unicast MPEG4/TCP streams from the Scene Recorder, the recordings are taken from the Scene Database.

Each client request is either directly served from the cache (*cache hit*), or a plug-in for the particular output format is invoked which transcodes the target stream using the given parameters (*cache miss*). Most plug-ins for the current implementation therefore utilize *FFmpeg* - an open source, cross-platform solution to record, convert, and stream audio and video [10]. The transcoded data is immediately delivered to the client, as soon as the plug-in has produced the first bytes. This enables a seamless video delivery to the user, who cannot tell whether the request is served from the cache or not.

For both client applications, Flash video is selected as the default output format. Each application therefore integrates its own *TMC Player*, which – after proper configuration - automatically takes over communication with the Scene Recorder or Scene Database in live stream and the recording requests, respectively. This player is based on the open source Flash player *JW Player* [11], which can be controlled by a comprehensive JavaScript API.

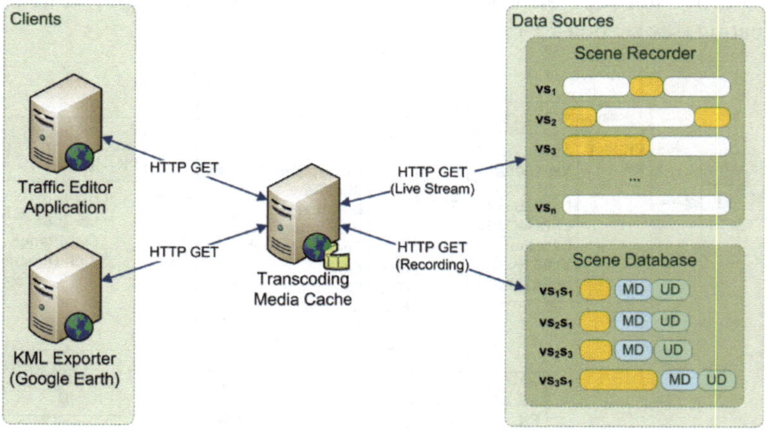

Fig. 6 Usage scenarios for the Transcoding Media Cache

2.5 Traffic Map Server

The verification of the accuracy of a traffic message also includes a verification of its location and spatial dissemination. To assist the traffic editors in this process, the Traffic Editor Application includes a visualization of the geometry of a traffic situation in terms of a web map (see section 2.6). This web map is provided by the *Traffic Map Server* (TMS).

The TMS consists of the three components (i) *Database Backend* with geometric and topological functions, (ii) *GeoServer*, and (iii) *GeoInterface* as the web front end. The Database Backend is again provided by a PostgreSQL server instance with PostGIS extension [12], which manages a database schema featuring all of the highways and motorways, sections, points, and cameras as geo-objects. Traffic messages are also imported from the Scene Database and their corresponding geometries are computed.

The GeoServer was adopted from the correspondent open source project GeoServer [13] and provides the intrinsic service for publishing geographic data. It implements an OGC compliant Web Feature Service, which is utilized to provide the GeoInterface with feature data about roads, points, messages, and cameras. This feature data is generated by querying the underlying Database Backend.

The GeoInterface is a web-based client to the GeoServer, which uses the OpenLayers library [14] for rendering the base layer in addition to the feature data from the GeoServer. Here the final composition of the web map takes place. The topmost layer is always comprised of traffic message and camera icons, which can be interactively selected to display details about the corresponding traffic messages or cameras, respectively.

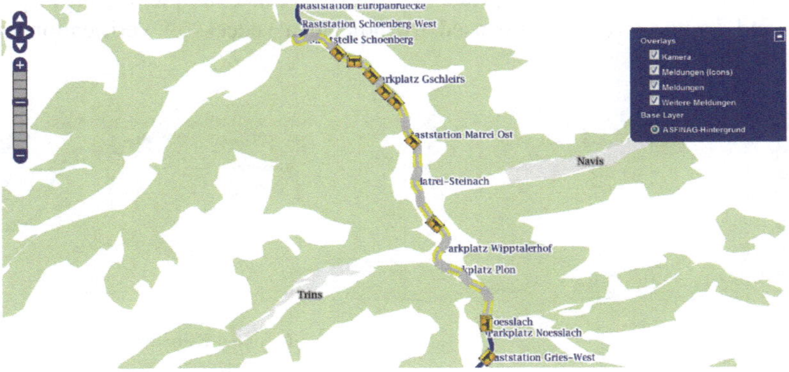

Fig. 7 An example traffic map for road works on the A13

An example traffic map illustrating the location and spatial expansion of a road works message on the A13 is shown in Fig. 7. The map is based on Open Layers and shows the background layer of ASFINAG, the traffic message layer, as well as the camera layer including all cameras with insight into the traffic situation area.

2.6 *Traffic Editor Application*

The web-based Traffic Editor Application is the final presentation and management application of our multimedia-centric traffic message verification system. It implements the top layer of the Scene Database (see section 2.3) and enables the traffic editors to verify traffic situations based on an integrated view of the details of a traffic message, its location and spatial dissemination in a web map, as well as the recordings and live streams provided by the Scene Recorder for the cameras with insight into the corresponding situation area. The Traffic Editor Application provides two views of the stored data: (i) a *List* view of all recent (or archived) traffic messages with filtering options to give the editors a quick overview about the events on Austria's roads, and (ii) a *Details* view describing a single traffic message.

Fig. 8 illustrates the *Details* view of this application based on the example of a LOS message of the *queuing* type and its corresponding situation on the A23. The location and geometry of this message are illustrated in the web map at the top of the page, including the cameras with insight into this location. On the left-hand side, details about the message are shown, which are relevant for message verification and scene publication. In this example, the queuing traffic message is related to a stationary traffic message for the same situation. Details about this related message are accessible in the area marked as (1). On the right-hand side, corresponding scene-recordings and live streams from the selected cameras can be requested in order to visually verify the current traffic message.

After the content and location verification process, the traffic editors may tag the traffic message with appropriate keywords. In this example, the tags *type verified* and *location verified* are used to indicate a successful verification of this

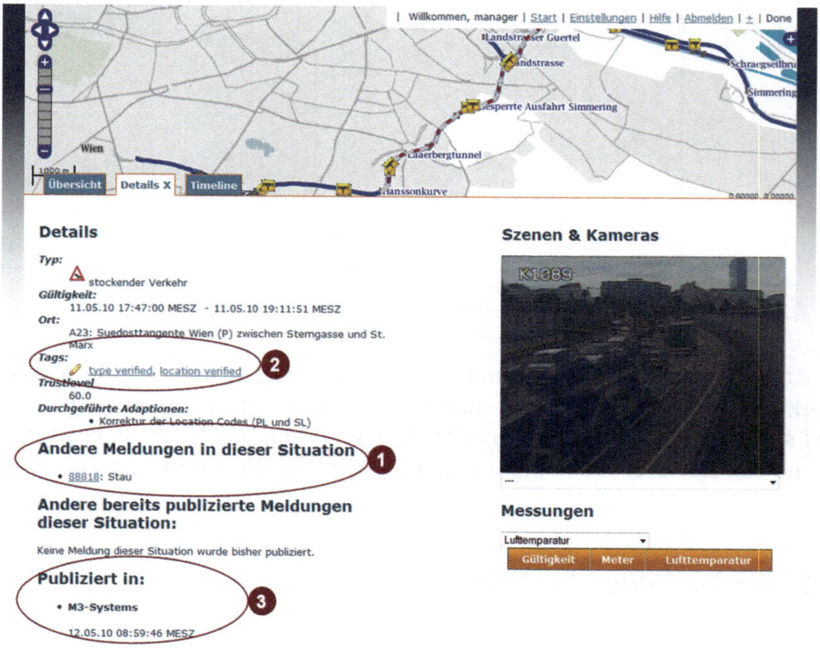

Fig. 8 Verification and publication of a LOS situation on the A23 (in German)

message (area (2)). Finally, in area (3) the traffic editors may also publish this traffic message including the related scene-recordings and available live streams for a particular user group, which may access these traffic scenes via the KML Exporter (see next section).

2.7 KML Exporter

The objective of the *KML Exporter* is to provide a more common access to the published traffic messages from the Traffic Editor Application for dedicated user groups. It exports the published traffic message scenes from the Scene Database as Keyhole Markup Language (KML) [15] documents, in order to enable their presentation in Google Earth. The geometry data of the traffic messages is thereby retrieved from the Web Feature Server of the Traffic Map Server. Scene-recordings and live camera streams can also be accessed via the TMC.

3 System Evaluations

The whole system has been deployed in the operational environment of ASFINAG since the beginning of 2009. It has been used by traffic editors at the national radio channel Hitradio Ö3, who evaluated, in particular, the benefits of the Traffic Editor Application for traffic message verification.

The main evaluation results were as follows: (i) The verification process for computer generated messages was improved considerably, since the accuracy of these messages was rather low. For example, LOS messages with level LS1 (stationary traffic) could be falsified in many cases. Since the sources of these messages are among the fastest, this verification information can be used by the editors for further decision-making regarding the corresponding traffic situations. (ii) In the case of slower message sources with a higher confidence level, the current verification process often comes too late. There are no longer any published traffic situations featuring peculiarities at the point of verification, although such peculiarities do exist at publication time. For example, in some cases, LOS messages with level LS1 were published, and our verification system has displayed level LS2 (queuing traffic). (iii) For certain traffic messages, the number of cameras selected by this system is too large, which prevents efficient verification of certain situations. Fig. 9 gives an example of this - the case of road maintenance (RMT) and traffic restriction (RES) messages input into the system in April 2010 by Austria's turnpike maintenance agencies (ATBMS). The x-axis represents internal message IDs, and the y-axis the number of selected cameras for these messages. The chart shows that it is not unlikely that a large number of cameras are relevant for recording one single traffic situation. In March 2010 road maintenance messages were generated, involving nearly 60 cameras. Similarly high numbers were achieved for level of service (LOS) messages involving certain tunnels.

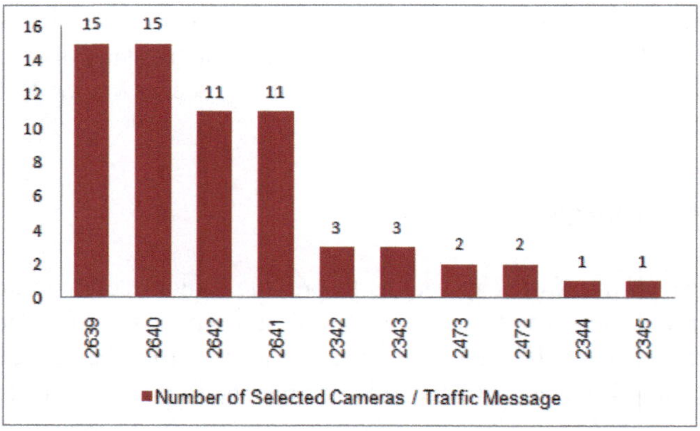

Fig. 9 Selected cameras for ATBMS traffic messages in April 2010

And finally, (iv) the most significant recordings or live streams should be presented in a more compact way – i.e. either in parallel or in a short summary. This is strongly related to the problem of determining the most significant cameras for verifying a traffic situation.

4 Conclusions and Future Work

This paper presented the architecture, services, and applications of the LOOK Platform – a system which provides an integrated traffic message verification approach by combining traffic message data with corresponding location maps and video streams from related surveillance cameras. The traffic message processing chain involves seven services and applications, which help to visually verify traffic messages and to publish them together with recorded scenes or camera live streams for certain user groups. For this purpose, the presented system interacts with the sensor and video subsections of the Austrian national operator of the highways and motorways, ASFINAG.

Evaluations of the system in an operational environment have shown that this system provides a great benefit by verifying the accuracy of computer generated LOS messages. Incorrect traffic messages can be detected earlier and corresponding new, improved messages can be published using high-confidence message sources. On the other hand, for slower but more accurate traffic messages sources, the verification process of the current system often comes too late. Furthermore, the camera selection and presentation approaches must be improved to enable a more efficient situation verification process. These shortcomings are currently being resolved in the successor system LOOK2, which begins traffic situation verification at the sensor level by utilizing sensor fusion techniques.

Acknowledgements. This work was supported by the Austrian Research Promotion Agency (FFG) under project 814244 (LOOK: A Platform for Traffic Content with Broad Access).

References

1. British Standards Institution, Road transport and traffic telematics, DATEX traffic and travel data dictionary (version 3.1a), BS DD ENV 13106:2000, 112 pages (2000)
2. Schulzrinne, H., Rao, A., Lanphier, R.: Real Time Streaming Protocol (RTSP), Internet Engineering Task Force, Request for Comments (RFC) 2326 (1998)
3. Jerzykowski, T.: MoRIS XML Protocol, Interface Description, Version 2.3.4, S-link AB, Sweden (2005)
4. NTCIP CCTV Working Group, Application Profile for XML Message Encoding and Transport in ITS Center-to-Center Communications (2008)
5. Ohm, J.R., Fisher, Y., Gill, A., Sunohara, O., Yagasaki, Y.: Information Technology – Coding of Audio-Visual Objects – Part 2: Visual, ISO/IEC 14496-2, 3rd edn. International Organization for Standardization (2003)
6. Schulzrinne, H., Casner, S., Frederick, R., Jacobson, V.: RTP: A Transport Protocol for Real-Time Applications, IETF, Request for Comments, RFC 1889 (1996)
7. Kikuchi, Y., Nomura, T., Fukunaga, S., Matsui, Y., Kimata, H.: RTP Payload Format for MPEG-4 Audio/Visual Streams, Internet Engineering Task Force, Request for Comments (RFC) 3016 (2000)
8. King, G., Bauer, C., Andersen, M.R., Bernard, E., Ebersole, S.: HIBERNATE – Relational Persistence for Idiomatic Java, Hibernate Reference Documentation 3.5.1. (2010)
9. ECMA International, ECMA Script Language Specification, Standard ECMA-262 (2009)
10. FFmpeg Project, General Documentation (2010), `http://www.ffmpeg.org/general.html` (accessed May 11, 2010)
11. Player, J.W.: Open Source Flash Video Player (2010), `http://www.longtailvideo.com/players/jw-flv-player/` (accessed May 11, 2010)
12. PostGIS Project, PostGIS Documentation, Version 1.5.1 (2010), `http://postgis.refractions.net/` (accessed May 11, 2010)
13. GeoServer Project, GeoServer User Documentation, Version 2.0. (2010), `http://docs.geoserver.org/stable/en/user/` (accessed May 11, 2010)
14. OpenLayers Project, OpenLayers: Free Maps for the Web (2010), `http://openlayers.org/` (accessed May 11, 2010)
15. Wilson, T.: OGC® KML, Version 2.2.0, Open Geospatial Consortium, OGC 07-147r2 (2008)
16. Kantowitz, B.H., Hanowski, J., Kantowitz, S.: Driver Acceptance of Unreliable Traffic Information in Familiar and Unfamiliar Settings. Human Factors 39 (1997)
17. Bogenberger, K., Hauschild, M.: QFCD – A Microscopic Model for Measuring the Individual Quality of Traffic Information. In: 16th ITS World Congress and Exhibition on Intelligent Transport Systems and Services, Stockholm, Sweden (2009)
18. Wischhof, L., Ebner, A., Rohling, H., Lott, M., Halfmann, R.: SOTIS – A Self-Organizing Traffic Information System. In: Proceedings of the 57th IEEE Vehicular Technology Conference (VTC 2003), Orlando, Florida, USA, pp. 2442–2446 (2003)
19. Wischhof, L., Ebner, A., Rohling, H., Lott, M., Halfmann, R.: Adaptive Broadcast for Travel and Traffic Information Distribution Based on Inter-Vehicle Communication. In: Proceedings of the IEEE Intelligent Vehicles Symposium, pp. 6–11 (2003)

Autonomous Multi-sensor Vehicle Classification for Traffic Monitoring

Horst Bischof, Martin Godec, Christian Leistner, Marcus Hennecke, Arnold Maier, Jürgen Wolf, Bernhard Rinner, and Andreas Starzacher

Abstract. The world will see a tremendous increase in the number of vehicles on the road in the near future. Future traffic monitoring systems will therefore play an important role in improving the throughput and safety of our roads. Current monitoring systems capture traffic data from a large sensory network. However, they require continuous human supervision or a significant amount of hand-labeled data for training and both are extremely expensive.

As part of a joint research project, we have investigated the scientific and technological foundations for future autonomous traffic monitoring systems. Autonomy is achieved by a novel combination of three approaches: self-learning and scene adaptive vision-based detection and classification, multi-sensor data fusion, and implementation on distributed embedded platforms.

In this paper we present our self-learning and co-training framework with the goal of significantly reducing the efforts required for manual training in data labeling and autonomously adapting the classifiers to changing scenarios. Our system consists of a robust visual online boosting classifier that allows for continuous learning. We also incorporate an audio sensor as an additional complementary source into the training process. We have implemented the framework on an embedded platform to support mobile and autonomous traffic monitoring. We have demonstrated this by detecting and classifying vehicles based on real-world traffic data captured on freeways.

Horst Bischof · Martin Godec · Christian Leistner
Graz University of Technology, Austria, Inffeldgasse 16/2, 8010 Graz
e-mail: {bischof,godec,leistner}@icg.tu-graz.ac.at

Marcus Hennecke · Arnold Maier . Jürgen Wolf
Efkon AG, Austria, Dietrich-Keller-Straße 20, 8074 Raaba
e-mail: {m.hennecke,a.maier,j.wolf}@efkon.com

Bernhard Rinner · Andreas Starzacher
Klagenfurt University, Austria, Lakeside B02b, 9020 Klagenfurt
e-mail: {bernhard.rinner,andreas.starzacher}@uni-klu.ac.at

J. Düh et al. (Eds.): Data and Mobility, AISC 81, pp. 15–26, 2010.
springerlink.com © Springer-Verlag Berlin Heidelberg 2010

1 Introduction

According to the US Department of Commerce and US Department of Energy [1], the world will observe a vast increase in the number of vehicles in near future. Due to the inherent limitations in developing and extending the current infrastructure of transportation systems, this tremendous increase will require more efficient use of roads and transportation means already in existence, the facilitation of traffic flow and, even more importantly, an increase in the safety of passengers. Thus, the next generation of traffic management systems will need more intelligent control systems to increase road safety as well as to decrease accidents and traffic congestion. These management systems require a large monitoring sensory network (typically visual sensors) and continuous human supervision, which is extremely expensive.

As a consequence, there has been a recent trend in research to develop *Intelligent Traffic Management Systems (ITMS)*. These ITMS aim to replace the supervisory manpower with intelligent agents. Such an ITMS consists of a network of local sensors that are mounted at various locations alongside/on/under/over existing roads. Each sensory device is capable of measuring some useful traffic parameters. Several different sensor devices have been applied to traffic monitoring in the past, e.g., laser/microwave radar, inductive loops, passive/active infrared sensors, magnetic devices, acoustic arrays and video cameras [2].

There is a strong interest in *Vision-based Traffic Measurement Systems (VTMS)* because they provide properties such as abstracted information content, non-intrusiveness, cost-effectiveness, portability, ease of maintenance and visualization capabilities. Currently, most traffic monitoring systems report only basic traffic parameters such as traffic flow and density. Intelligent VTMS are expected to provide more information about the actual state of the traffic. This information includes the occurrence of traffic accidents or jams, detection of driving violations or recognition of special vehicles (e.g., for electronic toll collection).

In this paper, we report on our joint investigations into future autonomous traffic monitoring systems. Autonomy is achieved by a novel combination of three approaches: firstly, vision-based detection and classification methods are augmented by self-learning and scene adaptation mechanisms, which significantly reduce the efforts required for manual configuration and calibration. Secondly, visual data is fused with data from other sensors such as acoustic and laser sensors. Multi-sensor data fusion helps to improve robustness and confidence. This extends the spatial and temporal coverage as well as reducing the ambiguity and uncertainty of the processed sensor data. Finally, the developed vision and fusion methods are implemented on a distributed embedded platform, which makes them more widely applicable and supports real-time operation.

The remainder of this paper is organized as follows:

Section 2 reports on related vision-based traffic monitoring systems. Section 3 introduces a framework for self-training visual classifiers exploiting audio sensors. Section 4 describes a generalization of this framework, moving towards autonomous co-training of audio and visual classifiers. Section 5 compares both frameworks and concludes this paper with a discussion about future work.

2 Related Work

Current monitoring systems capture (usually vision-based) traffic data from a large sensory network, but they require continuous human supervision, which is extremely expensive. Robustness and adaptivity are key challenges for intelligent traffic monitoring. Numerous sensors are installed at various locations. This diverse setting typically requires tedious sensor calibration and adapting the analysis algorithms to the observed scenes. This calibration and adaptation should be carried out with as little human intervention as possible. Most existing monitoring systems are designed to use a single sensor source (camera), but exploiting multiple different sensory information sources could improve the performance and reliability of the VTMS.

There are several examples of vision-based traffic monitoring systems such as [3], [4], [5] and [11]. Zhou et al. [6] propose an example-based algorithm to detect moving vehicles in a traffic monitoring scenario with changing environmental conditions. A similar approach is reflected in the work of Rodríguez et al. [7]. It describes a vision-based traffic monitoring system that is able to detect vehicles in real-time. The work aims to tackle some of the challenges in real-world applications such as shadows, day and night transitions, occlusions and slow traffic, achieving stable accuracy in these situations. Zhu et al. [8] propose an image-based automatic traffic monitoring system using a 2D spatio-temporal model. It provides the following functionalities: vehicle counting, vehicle speed estimation and classification by exploiting 3D measurements. In addition, some traffic monitoring systems exploit heterogeneous sensors and perform multi-sensor data fusion to improve the overall robustness and performance of the system. Klausner et al. [9] used acoustic and visual sensory devices for vehicle detection and classification. They performed sensor fusion at a feature and decision-level. Similarly, Kushwaha et al. [10] drew on acoustic and visual sensor information for vehicle tracking.

However, these systems, based on acoustic and visual classifiers, require continuous human supervision and a significant amount of hand-labeled data for training - both of which are cost-intensive tasks. Our approach focuses on autonomous acoustic and visual vehicle detection and classification. In contrast to the related approaches presented here, we propose an audio-supported visual self-learning framework with the goal of reducing the efforts required for manual configuration, and secondly, an audio-visual on-line co-training system that enables autonomous scene-specific adaptation. Both approaches significantly improve the classification performance compared to single sensor usage.

3 Audio-supported Visual Self-training

In the context of self-learning, a "trainer" guides the training process of a "trainee" with the goal of adapting to a specific scene without human intervention. The overall structure of our autonomous, audio-supported, visual, self-learning framework is depicted in Figure 1 [12]. The training process is shown on the left-hand

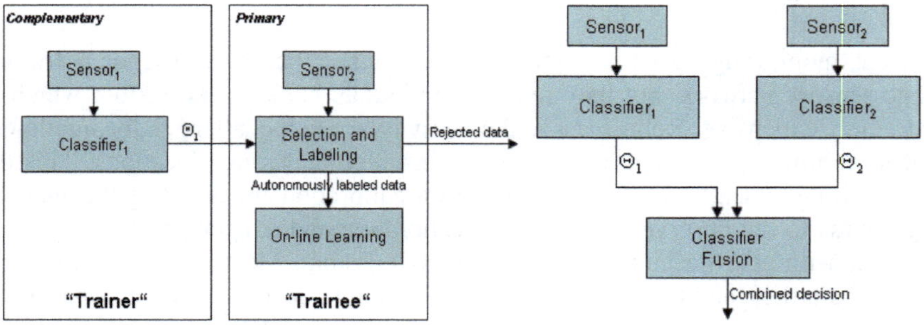

Fig. 1 Self-training framework: the training process (left) and the classification task (right)

side; it uses data from a primary and a complementary (audio) sensor source, respectively. The right-hand side presents the collaborative audio and video-based classification process.

In the training process, the acoustic and visual sensors synchronously capture data from the scene under consideration. The complementary sensor (audio) serves as a trainer for the online learning of the visual classifier (primary sensor). Initially, the audio classifier training is based on a small set of hand-labeled data. For the visual classifier, it is not a requirement to perform initial offline training. Each classification output from the complementary sensor's classifier is used to decide whether the classifier for the primary sensor s should be trained based on a specific sample or not. If confidence in the classification accuracy of the complementary classifier falls above a certain threshold, the primary classifier is trained based on this sample using the estimated class label assigned by the audio classifier. After the successful training of the primary classifier, it is used as an independent classifier for the collaborative classification task.

Similar to the work of Wu et al. [13] that trained an online classifier using a so-called "oracle" for pedestrian detection, we use an audio classifier as an autonomous instructor. Furthermore, we use robust online boosting as a visual classifier and incorporate the instructor in the final classification process to resolve ambiguities among vehicle classes.

3.1 Acoustic and Visual Classification

The basic structure of the audio classification system is depicted in Figure 2. After capturing the audio signal of passing vehicles along the road, the audio sample is partitioned into several blocks with a configurable block size. After this block-based partitioning, dedicated acoustic features are extracted for each block individually. These block features are then abstracted into a single feature vector using statistical merging. The feature vectors then serve as input for the appropriate classifier. The acoustic features are short-time energy, spectral bandwidth, spectral roll-off point, band-energy ratio values and a feature based on the cepstral

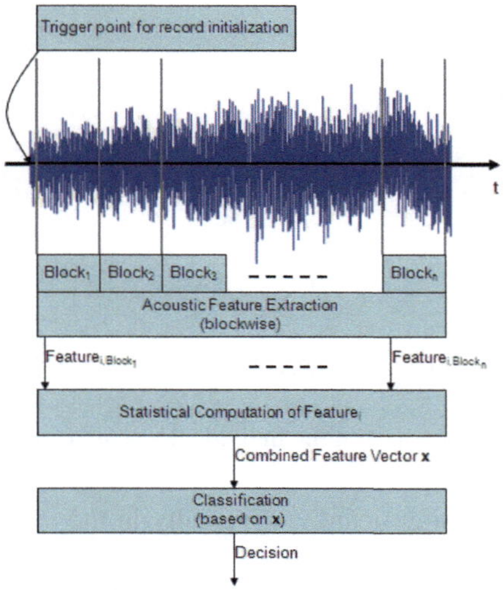

Fig. 2 Acoustic classification system; features are extracted from blocks of captured audio samples serving as input for the classifier.

analysis. An extensive evaluation of the selected acoustic features used for the acoustic classification is given in [14]. The acoustic classifier is based on the quadratic discriminant analysis technique.

The visual object detector allows for continuous learning without storing any training data by using online boosting for feature selection. Furthermore, a more robust loss function for the online boosting approach is applied [15]. As part of the training, we exploit the acoustic classifier in order to extract relevant training data from scene-specific video streams that are captured by a non-calibrated consumer camera. We train two distinct detectors, one for cars and one for trucks. In the classification stage, the acoustic classifier is incorporated into the considered objects of interest (collaborative classification using confidence-based sensor data fusion) in order to make the final classification.

3.2 Self-training Evaluation

This section describes the experimental evaluation conducted to demonstrate the feasibility of our self-learning approach and the improvements in classification accuracy compared to the performance of a visual sensor-only approach.

Figure 3 depicts the classification performance of the visual classifier after automatic training with the complementary sensor's classifier. The detectors for cars as well as trucks achieved high classification accuracy when applied to test scenes that only contain their training class, i.e., the detectors only discriminate the trained target class from the background of the scene (as shown on the left). If the visual detectors are applied to scenes containing both classes of vehicles, the

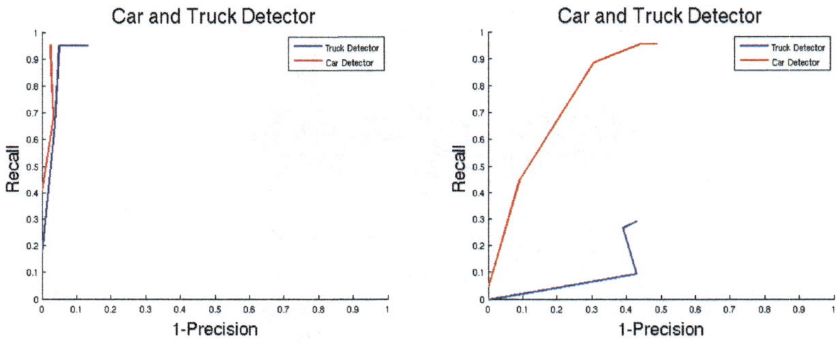

Fig. 3 Classification performance of the individual visual detectors for cars and trucks

performance deteriorates significantly, especially in the case of trucks (shown on the right).

However, the classification performance of the system significantly improves after incorporating the acoustic classifier into the classification process, performing a collaborative classification. The results of the improved joint classification performance of audio and video classifiers are shown in Figure 4.

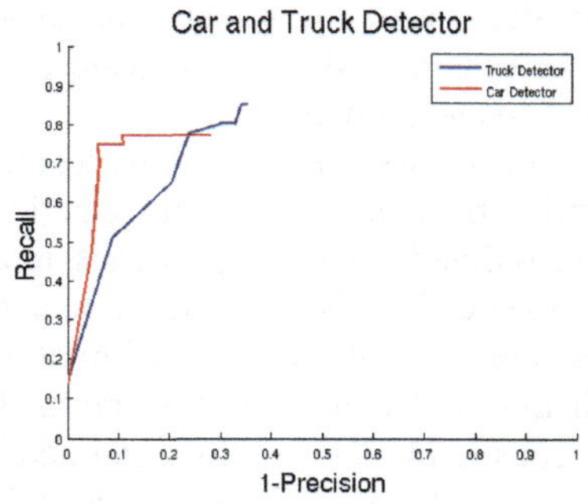

Fig. 4 The performance of the collaborative classification approach using audio and video information

Tables 1 and 2 present the overall performance of the self-learned visual classifier with the joint audio and visual classification. By comparing the F-Measure, which gives an impression of the overall performance, we can see the improvement of the combined classification.

Table 1 Classification performance using exclusively visual classifiers

Classifier	*Recall*	*Precision*	*F-Measure*
Truck	0.29	0.57	0.39
Car	0.95	0.51	0.67

Table 2 Classification performance using audio and visual classifier (joint classification)

Classifier	*Recall*	*Precision*	*F-Measure*
Truck	0.85	0.71	0.78
Car	0.77	0.77	0.77

4 Audio-visual Online Co-training

In this section we generalize our self-training approach and introduce autonomous, online, audio-video co-training [22]. Thus, the main idea is to initiate the training of (two) heterogeneous classifiers with very little labeled data and then to continue with the co-training of these classifiers based on a continuous stream of unlabeled data to yield scene-specific and highly adaptive classifiers.

As mentioned before, fusing audio and video information is not new and several classification applications take advantage of sensor fusion techniques. Our approach, however, differs in that we do not use a fusion strategy at the decision level, but rather we exploit the power of multiple sensors to perform robust autonomous learning. Christoudias et al. [16] describe a similar approach where audio-video co-training is performed for human gesture recognition. The authors use offline learning methods to exploit the entire training set at once. This requires the storage of all data but provides potential for optimization and typically achieves good results. In contrast, we use an online approach in order to learn from streaming data, such as video. Furthermore, we propose a multi-class co-training approach that enables discrimination between the different vehicle classes as well as against the background. One advantage of the online learning method is that we can reduce complexity due to scene-specific training. In addition, we do not need to store any data. Thus, our method can be implemented based on resource-constrained embedded systems as well. As in the self-training framework, we use non-calibrated consumer cameras and microphones.

4.1 Online Co-training System

Co-training was proposed by Blum et al. [17]. The original idea of co-training is to train two initial classifiers based on labeled data. Then, these classifiers only update one other using the unlabeled data set if one classifier is confident about a sample whereas the other is not. As shown in [17], the approach converges if two basic assumptions hold: first, the error rate of each classifier is low. Second, the views must be conditionally independent - something which was later relaxed in order to bridge the gap between theory and practice [18]. Thus, co-training can even be applied if the learners are slightly correlated.

In order to work, co-training needs two distinct views on the classification problem. The use of audio and video sensor information inherently offers real-world views that can be directly exploited through co-training.

We use online, random, naïve Bayes classifiers for the audio data. These classifiers are the simplest form of Bayesian networks [19]. An online multi-class boosting approach based on a logistic loss function is used for the visual classifier. These visual and acoustic classifiers are inherently multi-class, online compatible and robust to noise. We use the same acoustic features as in the self-training approach (also see [14]). As regards the visual features, we basically use Haar-like [20] and Haar-LBP features [21].

Our classifiers perform the processing task based on different scopes for capture time and object localization. Therefore, we have to synchronize both the evaluation and the co-training. We use a visual trigger that robustly delivers points in time where vehicles are present within a limited region in the visual view only. This trigger uses a robust block-based background model that handles camera shake due to changing illumination conditions, wind and vibrations. The visual trigger fires if the background model delivers a detection (based on a small part at the area within the region of interest that spans over all lanes in the current view). This is depicted in Figure 5.

Fig. 5 Examples of the visual triggers for cars and trucks

After firing, the visual classifier densely evaluates this region and finally delivers a potential detection of the vehicles present. The acoustic features are computed using recorded audio data of a few seconds. Both classifiers use their detections to train the other according to the co-training approach described above.

Finally, the visual classifier uses the location that obtains the highest confidence level according to the class label presented by the acoustic classifier.

4.2 Co-training Evaluation

The experiments are performed based on two different, real-world datasets recorded at several locations under varying weather conditions. The vehicles of interest are cars and trucks. The basic recording setup was as follows: a consumer microphone and camera were placed on a bridgeover to record passing vehicles. The audio recording parameters are an 8 kHz sampling rate, 16 bit resolution and mono format. The maximum recording duration was set to 5 seconds. The camera frame rate was set to approximately 20 Hz with a resolution of 640x480. The evaluation platform was a MICROSPACE EBX (*MSEBX945*) embedded computer board from *Digital-Logic AG*.

The baseline performance of the classifiers using only labeled data is presented in Table 3 (for comparison with the results obtained by using labeled and unlabeled data). As expected, the car classification performs well even with only 100 samples (top). However, the performance for the truck classifier is significantly lower (bottom).

Table 3 Initial classifier performance using different amounts of labeled data

Cars	*Recall*	*Precision*	*F-Measure*
100 Samples	0.96	0.94	0.95
150 Samples	0.98	0.97	0.98
200 Samples	0.96	0.96	0.96
Trucks	**Recall**	**Precision**	**F-Measure**
100 Samples	0.75	0.58	0.65
150 Samples	0.89	0.88	0.88
200 Samples	0.82	0.94	0.88

Our objective is to achieve at least the same performance with co-training but using only a few labeled samples for the initial training. Therefore, we use 100 labeled samples for the initial classifier and then perform co-training with unlabeled samples. Table 4 summarizes the results for different numbers of unlabeled data. Co-training clearly improves the performance of the initial classifiers for cars and trucks.

Table 4 Co-training performance using different amounts of unlabeled data (initial classifier trained with 100 labeled samples only)

Cars	*Recall*	*Precision*	*F-Measure*
Initial	0.96	0.94	0.95
100 Unlabeled	0.94	0.92	0.93
200 Unlabeled	0.96	0.93	0.94
Trucks	**Recall**	**Precision**	**F-Measure**
Initial	0.75	0.58	0.65
100 Unlabeled	0.86	0.85	0.86
200 Unlabeled	0.91	0.80	0.85

Another interesting aspect of co-training is whether an initially trained classifier can adapt to a new scene. To demonstrate this scene flexibly, we initially train our classifiers based on labeled positive samples from one scene. Then we adapt the classifiers by using unlabeled data from the new target scene. The two scenes have different viewing angles and object size. Table 5 summarizes the classifier performance based on 100 and 200 unlabeled samples posing a significant classification improvement for cars and trucks.

Table 5 Classifier performance for scene adaption based on100 and 200 unlabeled samples from the target scene

Cars	*Recall*	*Precision*	*F-Measure*
Initial	0.99	0.94	0.96
100 Unlabeled	0.99	0.97	0.98
200 Unlabeled	0.99	0.99	0.99
Trucks	**Recall**	**Precision**	**F-Measure**
Initial	0.48	0.46	0.47
100 Unlabeled	0.63	0.82	0.71
200 Unlabeled	0.70	0.71	0.70

Finally, we show that the audio-video co-training approach outperforms co-training based on visual classifiers only. Hence, two visual classifiers are trained using 100 samples. Then, co-training is performed on 200 unlabeled samples. The results are compared to the audio-video co-training approach. The results in Table 6 clearly indicate that our approach outperforms the visual-only approach.

Table 6 Co-training performance after 100 labeled and 200 unlabeled samples

Cars	*Recall*	*Precision*	*F-Measure*
Audio-Visual	0.96	0.93	0.94
Visual-Only	0.77	0.64	0.70
Trucks	**Recall**	**Precision**	**F-Measure**
Audio-Visual	0.86	0.85	0.86
Visual-Only	0.75	0.85	0.80

5 Conclusion

In this paper, we presented a self-learning and co-training framework for autonomous traffic monitoring systems exploiting acoustic and visual data sources. Powerful visual and acoustic classifiers exist, but in order to obtain high accuracy, these algorithms require a significant amount of hand-labeled data. Collecting this data is a tedious and cost-intensive task. In contrast, our framework (i) supports autonomous training to avoid manual labeling for every site, (ii) enables continuous online training to allow for varying scenario conditions such as weather and illumination changes, and (iii) is resource-effective to enable wide-spread use.

We have implemented the framework on an embedded platform and tested the performance with real traffic data recorded by uncalibrated consumer cameras and microphones. Our experiments show that our framework increases the classification ability by performing confidence-based sensor fusion and robustly adapts to new scenes without any additional manually labeled data.

Although we have demonstrated our framework for vehicle classification, self-learning and co-training are general concepts with a high potential for many applications. Future applications may include traffic monitoring, free-flow toll collection, and law enforcement. Thus, possibilities for future work include an extension to multi-class object classification, improving the detection of individual objects, i.e., trucks, and a comprehensive evaluation of accuracy, robustness, and scene adaptivity using larger test data sets. We are confident that our approach can serve as an important step toward versatile and cost-effective ITMS.

Acknowledgements. This work has been sponsored in part by the Austrian Research Promotion Agency under grant 813399.

References

1. Fuel cell vehicles: Race to new automotive future. Technical report, Technology Administration Report, US Department of Commerce (January 29, 2003)
2. Klein, L.A.: Sensor Technologies and Data Requirement for Intelligent Transportation Systems. Artech House, Boston (2001)
3. Kastrinaki, V., Zervakis, M., Kalaitzakis, K.: A survey of video processing techniques for traffic applications. Image and Vision Computing 21, 359–381 (2003)
4. Song, K.-T., Tai, J.–C.: Image-Based Traffic Monitoring With Shadow Suppression. Proceedings of the IEEE 95(2), 413–424 (2007)
5. Cucchiara, R., Piccardi, M., Mello, P.: Image Analysis and Rule-Based Reasoning for a Traffic Monitoring System. IEEE Transactions on Intelligent Transportation Systems 1(2), 119–130 (2000)
6. Zhou, L.J., Gao, D., Zhang, D.: Moving Vehicle Detection for Automatic Traffic Monitoring. IEEE Transaction on Vehicular Technology 56(1), 51–59 (2007)
7. Rodríguez, T., García, N.: An adaptive, real-time, traffic monitoring system. Machine Vision and Applications 18, 781–794 (2009)
8. Zhu, Z., Xu, G., Yang, B., Shi, D., Lin, X.: VISATRAM: A real-time vision system for automatic traffic monitoring. Image and Vision Computing 18, 781–794 (2000)
9. Klausner, A., Leistner, C., Tengg, A., Rinner, B.: An audio-visual sensor fusion approach for feature based vehicle identification. In: Proceedings of the 2007 IEEE International Conference on Advanced Video and Signal based Surveillance (AVSS 2007), London, UK, p. 6 (2007)
10. Kushwaha, M., Ohy, S., Amundson, I., Koutsoukos, X., Ledeczi, A.: Target Tracking in Heterogeneous Sensor Networks Using Audio and Video Sensor Fusion. In: IEEE International Conference on Multisensor Fusion and Integration for Intelligent Systems, Seoul, pp. 14–19 (2008)
11. Koutsia, A., Semertzidis, T., Dimitropoulos, K., Grammalidis, N., Georgouleas, K.: Proceedings of the Sixth International Workshop on Content-Based Multimedia Indexing, pp. 125–132 (2008)
12. Bischof, H., Godec, M., Leistner, C., Starzacher, A., Rinner, B.: Autonomous Audio-Supported Learning of Visual Classifiers for Traffic Monitoring. IEEE Intelligent Systems 25(3), 15–23 (2010)
13. Wu, B., Nevatia, R.: Improving part based object detection by unsupervised, online boosting. In: IEEE International Conference on Computer Vision and Pattern Recognition (CVPR 2007), pp. 1–8 (2007)

14. Starzacher, A., Rinner, B.: Single Sensor Acoustic Feature Extraction for Embedded Realtime Vehicle Classification. In: Proceedings of the Second International Workshop on Sensor Networks and Ambient Intelligence (SeNAmI 2009), pp. 378–383 (2009)
15. Leistner, C., Saffari, A., Roth, P.M., Bischof, H.: On Robustness of Online Boosting - A Competitive Study. In: 3rd IEEE On-line Learning for Computer Vision Workshop (ICCV 2009), vol. 1, pp. 246–252 (2009)
16. Christoudias, C., Urtasun, R., Kapoorz, A., Darrell, T.: Co-training with noisy perceptual observations. In: IEEE International Conference on Computer Vision and Pattern Recognition (CVPR 2009), Miami, FL, USA, pp. 2844–2851 (2009)
17. Blum, A., Mitchell, T.: Combining labeledand unlabeled data with co-training. In: Proceedings of COLT 1998, pp. 92–100 (1998)
18. Balcan, M.-F., Blum, A., Yang, K.: Co-training and expansion: Towards bridging theory and practice. In: Advances in Neural Information Processing Systems, pp. 89–96 (2004)
19. Geiger, D., Goldszmidt, M., Provan, G., Langley, P., Smyth, P.: Bayesian Network Classifiers. Machine Learning, 131–163 (1997)
20. Viola, P., Jones, M.J.: Rapid object detection using a boosted cascade of simple features. In: IEEE International Conference on Computer Vision and Pattern Recognition (CVPR 2001), vol. 1, pp. 511–518 (2001)
21. Roy, A., Marcel, S.: Haar local binary pattern feature for fast illumination invariant face detection. In: BMVC (2009)
22. Godec, M., Leistner, C., Bischof, H., Starzacher, A., Rinner, B.: Audio-Visual Co-Training for Vehicle Classification. In: Proceedings of the 7th IEEE International Conference on Advanced Video and Signal Based Surveillance (AVSS), Boston, MA, USA (August/September 2010)

Concepts for Modeling Drivers of Vehicles Using Control Theory

Martin Rudigier and Martin Horn

Abstract. Numerical simulation is a key technology for vehicle development. In vehicle dynamics simulations, the resulting dynamics heavily depend on the interaction of the vehicle, environment and human driver. Suitable mathematical models of human drivers for automotive applications are required in order to gain significant simulation results. The driver can be regarded as a controller for the system consisting of the vehicle and environment (road and traffic). In most vehicle dynamics simulations, the driver is modeled as a controller in an engineering sense. However, characteristics that describe the "human" behavior of a driver should be taken into account as well. In the first part of this article, a short overview of methods of modeling human drivers is given. Subsequently a driver model for vehicle dynamics simulation is presented. A set of basic and advanced use cases define the requirements for this driver model within a professional vehicle dynamics package (multi-body simulation). These use cases extend from "straight line acceleration" to complex driving maneuvers, such as the "3-point-turn", i.e. changing driving direction on a road that is narrower than the turning radius of the vehicle. The driver model is discussed based on exemplary simulation results. Finally a concept of how to adapt this model to real human behavior is presented.

1 Introduction

Over time, simulation has become an increasingly important tool for automotive development. The dynamics of the vehicle are not only determined by themselves, but instead are made up of the "vehicle-driver-environment" system. With open-loop maneuvers, the driver could be excluded for a certain amount of test scenarios, but for many other test scenarios, models of the human driver are necessary.

Martin Rudigier
Virtual Vehicle, Graz

Martin Horn
Institute for Smart System Technologies, Control and Measurement Systems,
University of Klagenfurt

J. Düh et al. (Eds.): Data and Mobility, AISC 81, pp. 27–38, 2010.
springerlink.com © Springer-Verlag Berlin Heidelberg 2010

Besides vehicle dynamics simulation, other areas of research also need models of the driver e.g. research into understanding the behavior of human drivers or traffic simulations. Vehicle safety research needs driver models for crash simulations, but these models describe the body, skeleton and muscles. These models are not addressed in this article. This paper deals with the driver as the operator of a vehicle. The focus is on the control aspect of the human driver and modeling this. In addition, other aspects of driving such as path-following and human factors - cognitive aspects, driving styles and attention – are also addressed.

2 Driver Models – An Overview

A large number of publications dealing with very different driver models exist. In this section, an overview and some sort of categorization is given. Firstly, the tasks of a driver and the characteristics of human drivers are analyzed. Then the driver models are categorized based on application. Jürgensohn subdivides the driver models in three classes [4]: driving models, driver models and evaluation models. In this classification, *driving models* are used to stimulate vehicle dynamics in simulations. *Driver models* reproduce human behavior. For *evaluation models,* the main goal is to benchmark the vehicle dynamic. This classification only addresses simulation models. Salvucci groups these models into one class and introduces a new class where other aspects are taken into account [9]:

- computational models that compute, simulate, and predict various aspects of driver behavior,
- conceptual models that help to understand the representational and procedural components of the driving task.

This paper deals with the first class of models mentioned by Salvucci; models of the second class are only covered briefly.

2.1 The Task of Driving

The task of driving combines a couple of basic tasks. These basic tasks change continuously. Various authors identify several classes of tasks. Bösch [1] ranks three tasks in a hierarchical structure with three levels,

- navigation level,
- maneuver level and
- control and stabilization level.

The navigation level deals with the task of planning which roads have to be used in order to travel from location A to location B. In the next level, the maneuver level, the *maneuver planning* of the vehicle motion within the area in sight is covered. *Maneuver planning* means the choice of speed and path. The control and stabilization level (*maneuver execution*) is the third level in this model. This level is mandatory for the execution of the maneuver and the stabilization of the vehicle.

Salvucci uses three classes of process to describe the task of driving: operational processes, that involve manipulating control inputs for stable driving, tactical processes that govern safe interactions with the environment and other vehicles and strategic processes for higher level reasoning and planning. Furthermore, secondary tasks are identified which can be related (using a navigation device) or unrelated (dialing a cell phone) to driving [9].

Depending on the subject of research, models include properties of different levels and secondary tasks. For the investigation of driving dynamics, only the control and stabilization level is needed. The other levels of the hierarchical model are prepared by the simulation engineer when he decides which test he wants to simulate. For research on active suspension systems and driver assistance systems, the maneuver level is also modeled. In motorsports, maneuver planning can also be used for lap simulations and setup variations [7]. Secondary tasks have been modeled for human factor research into things like side-tasking and distraction.

2.2 Characteristics of Human Drivers

According to MacAdam [5], various human characteristics have influence on vehicle operation while driving:

- information reception, perception and processing,
- neuromuscular dynamics with threshold, time delay and limitations,
- preview, anticipation,
- adaption / learning,
- planning capabilities (path and speed),
- experience,
- risk behavior
- portraying features: concentration, tiredness / stress and emotion.

For many applications, most of those characteristics are negligible. When modeling the human driver, the right choice of characteristics given by the application has to be made. To achieve this, it is necessary to know which characteristics influence the driver in the driving situation in question e.g. when following a path with a certain speed, time delays play an insignificant role, because if the situation is predictable, the ability of anticipation almost compensates for reaction time and other time delays.

Driving capability is initially constrained by biological characteristics, like information processing capacity and motor coordination. Training and experience enhance the knowledge and skill of human drivers. These factors limit the driving capability of drivers, but drivers are unable to exploit this potential all the time due to human factor variables. These variables include factors such as attitude, motivation, effort, fatigue, drowsiness, time-of-day, drugs, distraction, emotion and stress. Any of these may potentially cause the driver's driving capabilities o fall the limits [3]. For the major part of vehicle dynamic development, these considerations may not be useful, but for accident analysis and prevention they are important.

Human drivers have different characteristics; their level of aggression differs and their level of experience varies. Therefore a lot of driver models try to group the drivers together and develop some forms of "typical" driver. Typical attributes when selecting a driving style for a driver model are "inexperienced/experienced" and "dynamic/with caution".

2.3 Applications with Driver Models

All three parts of the overall vehicle-driver-environment system are subjects of research. These areas are used to group the applications with driver models for research on the vehicle, on the driver and on environment.

Driver Models for Research on the Vehicle

The interest of vehicle dynamic applications generally focuses on the vehicle. The simulation task is usually a predefined maneuver. The maneuver is in most cases defined by given pedal and steering angle values (open loop maneuver) or a given path and velocity (closed loop maneuver).

For this kind of application, the main demand on the driver model is to be able to track a predefined path and speed trajectory by operating the steering wheel, throttle and brake pedal. Many models are able to choose their velocity according to lateral dynamics and maximum accelerations. During vehicle component tests, the driving task is reduced to follow given path and speed profiles. For these tasks, simple controllers without any attempt to mimic human behavior are often used. The test scenarios require models of test drivers that are sufficiently complex to represent both the learnt patterns and correction elements. Driver models, which include some level of anticipation, prediction, an internal model and a degree of compensation capability, may be suitable for this purpose [6]. The designing of onboard controls that influence vehicle dynamics and stability raises the problem of the interference between the driver as controller and the onboard control, e.g. many driver models have a higher bandwidth than human drivers and compensate for disturbances faster than a human, and therefore prohibit onboard control from working properly.

For research on the vehicle, individual behavior is not of interest therefore different classes of drivers are used (driving skill, dynamic).

Driver Models for Research on the Driver

Models for research on the driver have to be much more 'human-like' than models for research on the vehicle. These models can be used to research the impact of driver assistance systems on driver and traffic safety and for research on side-tasking e.g. dialing a cell phone [10]. Models for identification of driving styles belong to this class as well. These identified models can be used in driver assistance systems to adapt the system to the driver.

The models can be divided into three groups based on the modeling techniques: models with neural networks (NN), hybrid models and cognitive models.

Neural networks are used to represent adaptive human control behavior. In addition to simulating characteristic drivers, NN models are also used for identification or the continuous updating of driving styles.

Hybrid models combine different modeling methods to represent complexity and multi-layered human behavior. They are built up using sub-models on a psychological and physiological basis and combine rule-based models and controllers. An example of highway traffic simulation is presented in [2]. This is a hierarchical model comprising the 4 layers *decision making, task planning, maneuver* and *action*. In the first layer, the driver selects (rule-based, IF-THEN) a suitable task to be performed (normal traveling, overtaking, collision avoidance ...). The *task planning* layer decides which maneuver is carried out at what time to accomplish the selected task, based on the current traffic situation. A maneuver is composed of a sequence of basic actions. The *action layer* includes a few basic actions which are controllers e.g. a gap controller. For avoiding hazardous situations, e.g. a collision, a danger observer is set up based on fuzzy logic. Then the *decision making* layer may interrupt the current task and select the avoidance task.

Cognitive driver models are developed in a general cognitive architecture: a framework for specifying computational behavioral models of human cognitive performance. This architecture embodies both the abilities and constraints of the human system – for instance, abilities such as memory storage and recall, perception, and motor action; and constraints such as memory decay and limited motor performance [9]. The modules usually contain sub-models of perception, decision-making and neuromuscular activation. Such models can also be used to research driving capabilities during side-tasking situations. [10]

Driver Models for Research on the Environment

The main focus of research in the field of environment is traffic and the interaction with other vehicles. Traffic can be modeled according to a macroscopic view as traffic flow, where no individual vehicles are modeled, or based on a microscopic view where single vehicles are observed and individual behavior forms part of the research. These kinds of models have been used to investigate applications like adaptive cruise control (ACC) and their impact on traffic flow. Traffic simulations are also used to produce realistic and humanlike traffic for driving simulators. Models for the reconstruction of accidents and accident analysis/prevention also belong to this group, although they can partly be assigned to the second mentioned group, models for research on the driver.

After this short overview of driver modeling, sections 2 and 3 will concentrate on our research area, the development and simulation of vehicles.

3 A Driver Model for Multi-body Simulation

In this chapter, a driver model is presented which is developed for use in multi-body simulations (MBS). A set of use cases defines the requirements of the driver model. The set consists of some basic use cases and a couple of advanced use cases. The use cases contain maneuvers like "straight line acceleration", "steady state cornering", "drive in reverse" and "driving on rough road". Each use case is a sequence of different driving tasks. These driving tasks can be defined in different ways, e.g. follow a given velocity in the first task and follow a given acceleration in a second task. This feature allows the user to handle a wide spectrum of different driving scenarios, but leads to the use of different controllers during a scenario. It is important that the driver is able to switch between these controllers without producing additional disturbances.

In addition to control engineering and human behavior aspects, the MBS environment places further requirements on the driver model, which are continuous integration, multiple control variables (acceleration and velocity for longitudinal control) for the longitudinal and lateral driver, continuous switching and easy enhancement.

3.1 Model

The requirements outlined above lead to the structure shown in figure 1. The driver model is divided into a speed controller, a steering controller and a gearbox controller.

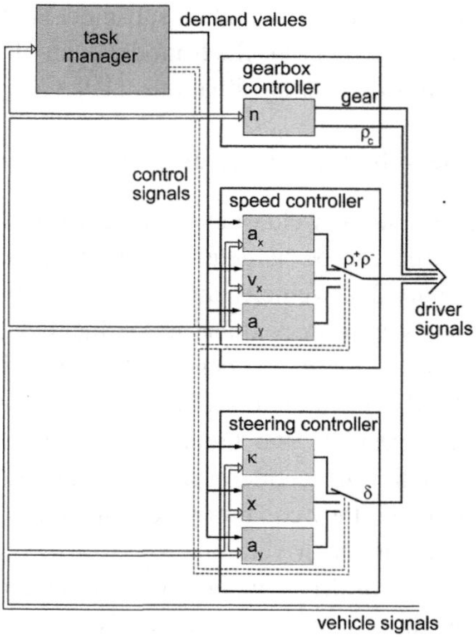

Fig. 1 Structure of the driver model [8]

The steering controller is capable of dealing with maneuver specifications defined by curvature, lateral acceleration or a path in Cartesian coordinates. For easy enhancement, a cascaded structure was chosen. Figure 2 shows the sub-controllers of the steering controller. When the maneuver is defined in the form of a position trajectory, the module position control generates a demanded curvature for the inner loop of the cascade, which is the curvature control.

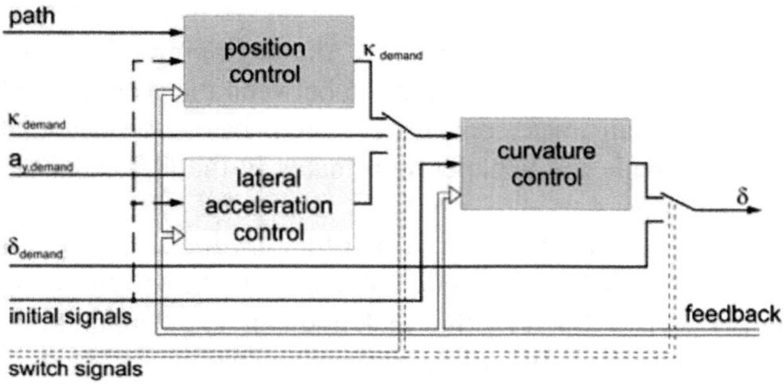

Fig. 2 Cascade structure of the lateral controller [8]

The demanded curvature is taken from the curvature of the circle defined through the actual position and direction of the velocity of the vehicle and the preview point $\mathbf{x}_{preview}$ (figure 3).

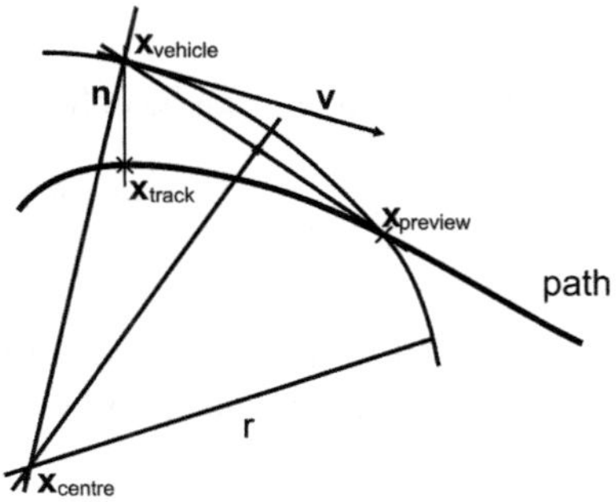

Fig. 3 Principle of the steering controller - position

The centre of this circle is derived as the intersection of two lines. The first line is formed by the position of the vehicle $x_{vehicle}$ and the direction $\mathbf{n}$, perpendicular to the direction of the velocity $\mathbf{v}$ and the second by the center point between the vehicle position $x_{vehicle}$ and the preview position $x_{preview}$, perpendicular to the vector from the vehicle position to the preview position. With

$$\tilde{\mathbf{x}}^{Tr} := (\mathbf{x}^{Tr}_{vehicle} - \mathbf{x}^{Tr}_{preview}) = \begin{pmatrix} \tilde{x}_1 \\ \tilde{x}_2 \end{pmatrix}, \qquad (1)$$

the formula for the demanded curvature κ_{demand} can be derived as

$$\kappa_{demand} = \frac{1}{r} = -2\frac{(n_1\tilde{x}_1 + n_2\tilde{x}_2)}{(\tilde{x}_1^{\,2} + \tilde{x}_2^{\,2})}. \tag{2}$$

The preview distance $s_{preview}$ consists of a constant distance $s_{preview,0}$, a velocity dependent part $\mathbf{v}$ times $t_{preview}$ and the actual position error s_2. The obtained curvature is used as input to the curvature controller, which is described in [8].

The track is defined in the track coordinate system. It has to start at the point $\mathbf{x}^{Tr} = (0,0,0)^T$ and with the direction $\theta^{Tr} = 0$. At the beginning of each task, when the position control is selected, a transformation between the global coordinate system and the actual Cartesian track coordinate system has to be calculated. The task controller has to provide the vehicle coordinates in the Cartesian task coordinate system.

The track can be seen as the description of a curvilinear coordinate system: the coordinates are s_1, the distance travelled along the path, and s_2, the distance to the path. The angle $\theta(s)$ is defined by the track and can be calculated in advance. It is calculated during the preparation of the maneuver data.

No constant transformation exists between the Cartesian and the curvilinear coordinates system. Both coordinates can be derived by integration, but the difference between these two coordinates increases due to numerical errors. With the assumption that s_1 and s_2 can only be approximated, an approximated value of $\mathbf{x}_{vehicle}$ is calculated by

$$\hat{\mathbf{x}}^{Tr}_{vehcile} = \mathbf{x}^{Tr}_{Tr}(\hat{s}_1) + \hat{s}_2\begin{pmatrix} -\sin\theta^{Tr} \\ \cos\theta^{Tr} \end{pmatrix}. \tag{3}$$

With this value, an approximation error $\mathbf{e}^{Tr}$ can be defined as:

$$\mathbf{e}^{Tr} =: \mathbf{x}^{Tr}_{vehicle} - \hat{\mathbf{x}}^{Tr}_{vehicle} = \mathbf{x}^{Tr}_{vehicle} - \left(\mathbf{x}^{Tr}_{tr} + \hat{s}_2\begin{pmatrix} -\sin\theta^{Tr} \\ \cos\theta^{Tr} \end{pmatrix}\right). \tag{4}$$

This error has to be transformed from the track coordinate system to the curvilinear system and is added to integration in the curvilinear coordinate system:

$$\dot{\hat{\mathbf{s}}}^{Tr} = v_x^V \cdot \begin{pmatrix} \cos(\psi^{Tr} - \theta^{Tr}(s_1)) \\ \sin(\psi^{Tr} - \theta^{Tr}(s_1)) \end{pmatrix} + v_y \cdot \begin{pmatrix} -\sin(\psi^{Tr} - \theta^{Tr}(s_1)) \\ \cos(\psi^{Tr} - \theta^{Tr}(s_1)) \end{pmatrix} + k_{p,ob} \cdot \mathbf{D}(\theta^{Tr}(s_1)) \cdot \mathbf{e}^{Tr}. \tag{5}$$

By choosing an adequate value of $k_{p,ob}$ the error is kept small.

3.2 Simulation Results

The driver model was first tested with a two-track-model developed in Matlab/Simulink, and then with a full vehicle model of a SUV, modeled with the MBS software program Altair MotionSolve. The use cases were also the test scenarios.

In this section, some results of the use case for "steady state cornering" are presented.

This use case is divided into two driving tasks. The first task is *"Initialization and reaching steady-state condition"*. The vehicle is driven by an open-loop control at a constant speed on a circle. Pedals and steering angle are chosen to meet the velocities and accelerations of the second driving task. After ten seconds, the task controller switches to the second driving task, "*Gain control and deal with disturbances*" and activates the necessary controllers, the "steering control - position" and the "speed control - velocity". The path was calculated in order to reach the desired radius prior to the simulation start. At fifteen seconds, a gust of side wind takes effect on the vehicle.

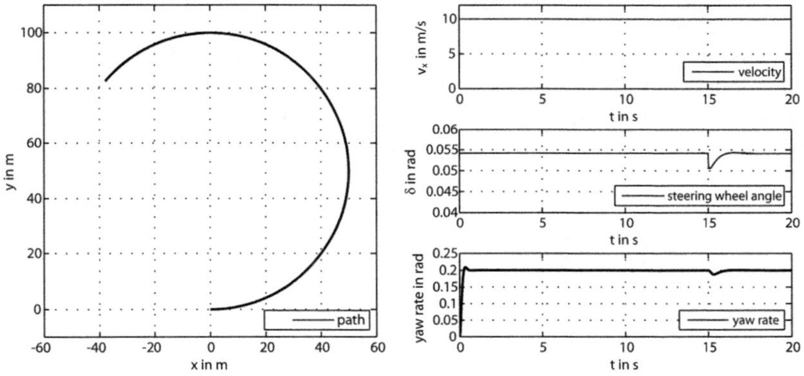

Fig. 4 Simulation results

Figure 4 displays graphs of some vehicle states during simulation. On the left side, the path of the vehicle is shown. In the right upper corner, the velocity is depicted. In the middle of the right side, the steering wheel angle can be seen. The right lower graph shows the yaw rate.

The vehicle model requires two seconds to reach the steady state. After ten seconds, the closed loop controller gains control. This causes a small disturbance in all 3 signals on the left side, but the disturbance is too small to be identified in figure 4. After 15 seconds, a disturbance can be seen, when the gust of wind affects the vehicle.

4 Getting Closer to Human Behavior – A Perspective

When considering the overview of driver models presented in section 1 and other more complete summaries of driver modeling [5,6], one may conclude that suitable models for almost every application exist. However, when simulating vehicle dynamics, the engineer faces several problems. In most cases, driver models only solve the scenario for which they are designed with good results. Finding parameters for other scenarios is often very difficult and time consuming. We want to model a driver for our field of interest, a vehicle dynamics simulation suitable for a wide area of test scenarios. The driver model introduced in section 2 lacks human behavior. The cascaded structure can be easily extended based on models of desired features.

Figure 5 gives an overview over the structure of our plans. Step by step, the model should be extended. The shown structure is not complete yet. Blocks with a grey background mark existing sub-models including the position control presented in section 2. The blocks with a white background and black frames represent our next steps and the blocks with the grey frames will be followed up later. The main modules of the model are perception, navigation level, maneuver level and control/stabilization level. Additional modules for driving styles and mental vehicle models are envisioned.

The model navigation level contains the definition of the driving task and the module model control. The engineer defines the driving task during preprocessing. The task of deciding which roads are used is not part of this driver model. The model control activates the necessary controllers for fulfilling the driving task.

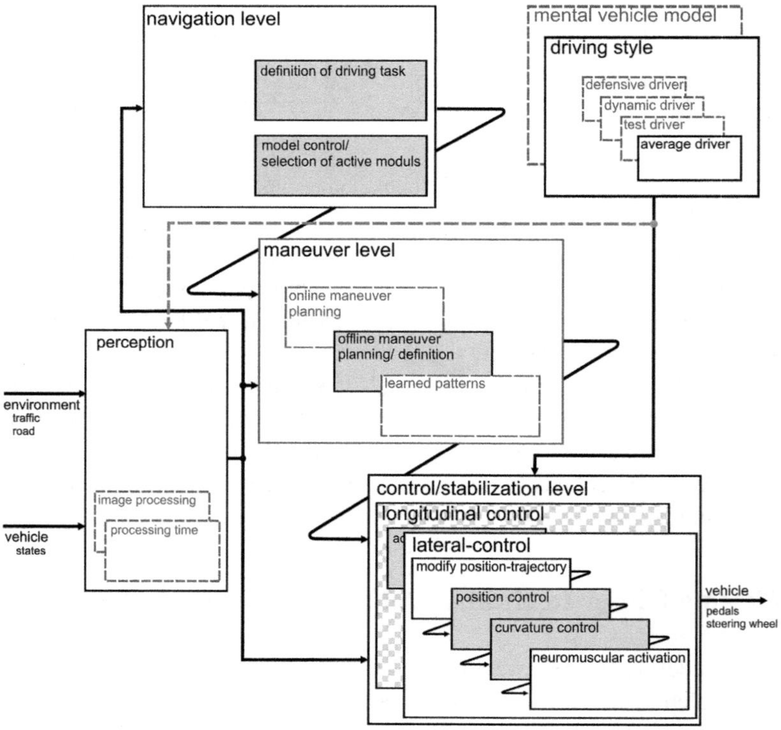

Fig. 5 Concept for the further development of the driver model

The module for 'maneuver planning' is responsible for planning the path and speed. The path and speed trajectory are defined by the user during preprocessing. Later this module should be enhanced, e. g. with a module for online maneuver planning. The focus of the development is the maneuver execution. In the first step, the existing cascade consisting of the position control and curvature control will be enhanced with a steering angle controller. The input to this controller is formed by the steering angle and the output is the steering force. This controller should model the dynamic of the neuromuscular activation and the movement of the limbs. The major objective of this controller is not optimal control, but modeling human characteristics. A similar controller has to be added to the longitudinal control. The inputs for the longitudinal control are the pedals and the outputs are the force on the pedals.

The second step is the development of a module for modification of vehicle accelerations. The maneuver specification given by the user has to be adjusted to match the acceleration typical of human drivers, potentially by varying the path and acceleration trajectories. The parameters of the driver model are chosen to mimic an average driver. In subsequent steps, parameters for test drivers, and different driving styles should be identified.

With finishing these steps, the area of "driving robots" should be left and the model will get closer to human driver models for vehicle dynamic simulations.

Acknowledgements. The authors would like to acknowledge the financial support of the "COMET K2 - Competence Centres for Excellent Technologies Programme" of the Austrian Federal Ministry for Transport, Innovation and Technology (BMVIT), the Austrian Federal Ministry of Economy, Family and Youth (BMWFJ), the Austrian Research Promotion Agency (FFG), the Province of Styria and the Styrian Business Promotion Agency (SFG).

We would furthermore like to express our thanks to our supporting industrial and scientific project partners, namely Altair Engineering, University of Klagenfurt and to Graz University of Technology.

References

1. Bösch, P.: Der Fahrer als Regler, PhD thesis, Vienna University of Technology (1991) (in German)
2. Cheng, B., Fuijoka, T.: A Hierarchical driver model. In: Proceedings of IEEE Conference on Intelligent Transportation Systems, pp. 960–965 (1997)
3. Fuller, R.: Towards a general theory of driver behavior. Accident Analysis and Prevention 37, 461–472 (2005)
4. Jürgensohn, T., Kupschick, S., Sölter, K.: Chucholowski, Virtuelle Testfahrer, Der Fahrer im 21, pp. 35–52. VDI Verlag, Jahrhundert (2009)
5. MacAdam, C.C.: Understanding and Modeling the Human Driver. Vehicle System Dynamics 40, 101–134 (2003)
6. Plöchl, M., Edelmann, J.: Driver models in automobile dynamics application. Vehicle System Dynamics 45, 699–741 (2007)

7. Prettenthaler, C., Rudigier, M., Horn, M.: Optimale Fahrzeugführung für Rennsport-simulationen. In: SSRP 2007, pp. 77–116 (2007)
8. Rudigier, M., Prettenthaler, C., Horn, M.: Betrachtung der Reglerumschaltung in einem Fahrermodell für Fahrzeugdynamiksimulationen. Steirisches Seminar über Regelungstechnik und Prozessautomatisierung 16, 40–61 (2009)
9. Salvucci, D.: Modeling driver distraction from cognitive tasks. In: Proceedings of the 24th Annual Conference of the Cognitive Science Society, pp. 792–797 (2002)
10. Salvucci, D.: Modeling driver behavior in a cognitive architecture. Human Factors 48, 362–380 (2006)

COOPERS: Driver Acceptance Assessment of Cooperative Services

Results from the Field Test in Austria

Doris Bankosegger

Abstract. The underlying document deals with the assessment of user acceptance in the field of Cooperative systems and services. The assessment of user acceptance is considered to be a key element involved in the prediction of the future market penetration of innovations. This paper focuses on the user acceptance of Cooperative services, which represent one key aspect of an EU-funded Integrated Project named COOPERS (Co-operative systems for intelligent Road Safety). Within this project, several field tests in Europe have been performed to evaluate user acceptance and driver behavior. The underlying paper focuses on the results of a field test performed in Austria/Innsbruck. A sample of 48 drivers was tested in January/February 2010 in cooperation with the Austrian Road operator, Asfinag. This paper shows the results of this user acceptance test and will give an indication of test driver expectations and experiences with an innovate Cooperative System, that provides the driver with safety-critical and convenience services.

1 Introduction

In technology acceptance literature, the terms 'acceptance', 'adoption', and 'attitude' are often used in various ways. A conceptual demarcation of the different terms is therefore required with a clear definition for technology/user acceptance. Figure 1 shows the different stages ranging from attitude towards acceptance on a chronological line [1].

Technology or user acceptance is thus defined as the degree to which individual users will use a given system when usage is voluntary or discretionary [2]. The acceptance of a product or service refers to the continued usage of it. It can be measured by two parameter values: frequency of use and intensity of use.

Doris Bankosegger
High Tech Marketing, Lothringerstraße 14/6, 1030 Vienna
e-mail: db@hitec.at

J. Düh et al. (Eds.): Data and Mobility, AISC 81, pp. 39–47, 2010.
springerlink.com © Springer-Verlag Berlin Heidelberg 2010

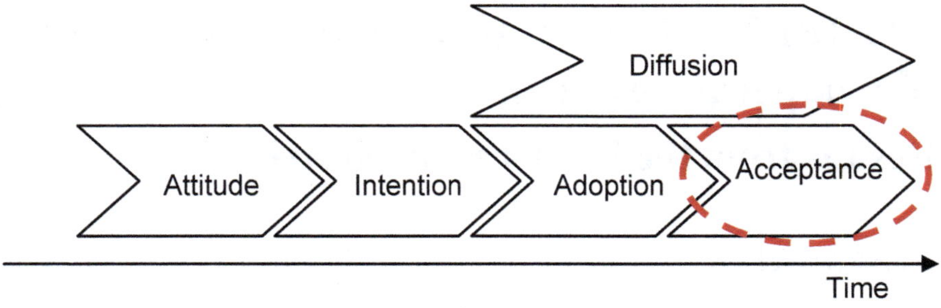

Fig. 1 From Attitude towards Acceptance [1]

Technology Acceptance is one of the most reliable predictors of the current system usage [3]. It is important to identify and measure users' perceptions of the system in order to predict the system usage and the adopted level of technological sophistication [4]. Where heavy investments are required before systems can be implemented (as it is the case with infrastructure and car manufacturing), it becomes a necessity to take user-acceptance considerations into account early in the development phases, thus setting the best preconditions for the subsequent wide-ranging adoption and intensive use of the system.

The following section gives an overview of selected methods for assessing user or technology acceptance.

2 Measuring Technology Acceptance

2.1 Methods

2.1.1 Unified Theory of Acceptance and Use of Technology (UTAUT)

A standard concept of user acceptance modeling widely adopted in information technology literature is the Unified Theory of Acceptance and Use of Technology (UTAUT) [5]. It combines several theoretical models, which try to explain acceptance and use of technology, including: 1. Theory of Reasoned Action [6], 2. Technology Acceptance Model (TAM) as well as TAM2 [7,5], 3. Motivational Model [8, 9], 4. Theory of Planned Behaviour (TPB) [10], Combined TAM-TPB [11], 6. Model of PC Utilization [12,13], 7. Innovation Diffusion Theory [14], and 8. Social Cognitive Theory [15]. Venkatesh, Morris, Davis and Davis [5] developed this unified model of technology acceptance. They found that UTAUT accounts for up to 70% of variance (adjusted R^2) in usage intentions. UTAUT is praised as a definitive model that synthesizes what is known and provides a foundation to guide future research in the area of technology acceptance [5, 16].

One of the most important basic theories of UTAUT, the TAM concept, is a widely used approach to assess information system usage. TAM's measurement qualities have also been validated inter-culturally in more than 100 studies. It translates the vast concept of prototype evaluation into well-defined research questions with recognized statistical benchmarks that can be answered in valid and reliable way.

Like in TAM, the key element for service acceptance in UTAUT is the extent to which the service is actually used. UTAUT is based on five indicators of user service acceptance:

- Performance Expectancy
- Effort Expectancy
- Social Influence
- Facilitating Conditions
- Behavioral Intention
- Usage Behavior

Gender, Age, Experience, and Voluntariness of Use are hypothesized to be key moderators in the suggested model. Furthermore, Playfulness, Expectation Confirmation, Satisfaction, Stress and Enjoyment are added as specific constructs to be regarded in the course of the COOPERS project. This model theorizes that four major constructs play a significant role as direct determinants of user acceptance and user behavior: performance expectancy, effort expectancy, social influence, and facilitating conditions [5]. These constructs can be tested by seven point Likert scales.

2.1.2 Rating Scale by Van der Laan

This method represents an adequate methodological approach to assessing the acceptance of advanced transport telematics. It is a single rating scale according to which drivers rate new technologies. As part of this rating scale, acceptance of a system is operationalized as a direct attitude towards that system. Attitudes are defined as predispositions to respond or tendencies in terms of "approach/avoidance" or "favorable/unfavorable". Attitudes are allocated to a basic bipolar continuum with a neutral or zero reference point, implying that they have both direction and intensity. The van der Laan rating scale consists of a set of nine scales. Individual items scores range from –2 to +2. The item numbers 3, 6 and 8 are mirrored, compared to the other items. The following table gives an illustration of the method.

Table 1 Rating scale by van der Laan

My judgments of the (…) system are….(please tick a box on every line)			
1	Useful	- - - - -	Useless
2	Pleasant	- - - - -	Unpleasant
3	Bad	- - - - -	Good
4	Nice	- - - - -	Annoying
5	Effective	- - - - -	Superfluous
6	Irritating	- - - - -	Likeable
7	Assisting	- - - - -	Worthless
8	Undesirable	- - - - -	Desirable
9	Raising alertness	- - - - -	Sleep alertness

3 COOPERS' User Acceptance Assessment

3.1 COOPERS Project

COOPERS is an EC FP6 funded IP, dealing with the application of real time data communication between the infrastructure and vehicle to exchange safety related information (e.g. speed, road conditions, local traffic information, etc.) with respect to the actual conditions of a certain road segment. Services defined in COOPERS include:

S1: Accident/incident warning (including wrong-way driver information)
S2: Weather condition warning (heavy rain, fog, hail, black ice, snow, heavy winds)
S3: Roadwork information
S4: Lane utilization information (lane banning, keeping in lane, auxiliary lane accessibility)
S5: In-vehicle variable speed limit information
S6: Traffic congestion warning
S7: ISA with infrastructure link
S8: International service handover
S9: Road charging to influence demand
S10: Estimated journey time (route navigation)
S11: Recommended next link (route navigation)
S12: Map information check to inform of current updates for digital maps (Route navigation)

Drivers receive these messages on an on-board unit similar to a navigation system in their vehicle.

3.2 Methodological Approach

Within the COOPERS Project, technology acceptance in addition to driver behavior has been tested on different test sites across Europe and in a simulator study. In the field test in Innsbruck/Austria which took place in January/ February 2010, 48 drivers participated. These participants were selected taking into account requirements of active User Involvement. The selected persons have Lead User Characteristics. Each test drive consisted of two actual tours, one tour *without using* a COOPERS system, one tour *using* a COOPERS system. The purpose of these two tours, which took place on the same highway section, was to be able to compare measured data when driving with the COOPERS system at a comparison level (measured when driving without the system).

The test procedure is shown in the following figure. It is split into two different stages, firstly, the preparatory stage, and secondly, the actual test. The preparatory stage is used to select suitable candidates for the test and for the test person to

Stage	Action	User Group 1	User Group 2
Preparatory Stage	User Selection	Driver Selection with Filter Questionnaire	
	Pre-test	Driving without and with the provisional fitted physiological measurement equipment	
Actual Test	Questionnaire I	Pre-questionnaire	Pre-questionnaire
	Diving I	Without Services	With Services
	Break		
	Driving II	With Services + GD	Without Services + GD
	Questionnaire II	Post-questionnaire	Post-questionnaire
	In-depth interviews	With selected drivers	

Fig. 2 Study layout for the field test and simulator

become familiar with the test vehicles as well as the measurement equipment in order to minimize these influences.

The actual test starts with the completion of a pre-questionnaire (user acceptance questionnaire). After the fitting and calibration of the PME, the driver goes for the first test drive, which lasts for about 30 minutes. After the re-calibration, the driver goes on the test drive for a second time. In order to balance the influence of the information system, half of the test drivers drive initially without and then with the services and the other half in the reverse order. After the second test drive, a post-questionnaire (user acceptance and driver behavior questionnaire) needs to be completed by the users. Selected drivers are interviewed in-depth after completion of the whole testing procedure.

4 Results of the Field Test in Innsbruck

As discussed in the methodology section, perceived usefulness and perceived ease of use of a system are the strongest predictors of actual system use and thus user acceptance. For this reason, the measures of the COOPERS field test for perceived usefulness and perceived ease of use as well as the intention to use are described here in more detail.

Figure 3 shows the results for the measure “perceived usefulness”. It depicts the single questions that were asked of the test drivers and shows the Likert-scale, where respondents are asked to rank their answers in a continuum between “strongly agree” and “strongly disagree”. The figure shows that the test persons had very positive expectations of the COOPERS system already before they had actually experienced it – the blue line indicates this expectation regarding the usefulness of the system as stated by the test drivers in the pre-questionnaire. The red line shows the average rating of the respondents after experiencing the COOPERS system. The figure shows, that the actual COOPERS system experience outperformed the test drivers' expectations: on average they found the system even more useful during driving than expected, that the system increases driving safety more than expected and that they gain improved information about the traffic situation. In the case of the questions "using the system I can move from A to B more quickly", "using the system I will enjoy improved driving convenience" and "using the system enables me to accomplish driving tasks", the system did not outdo the test drivers' expectations.

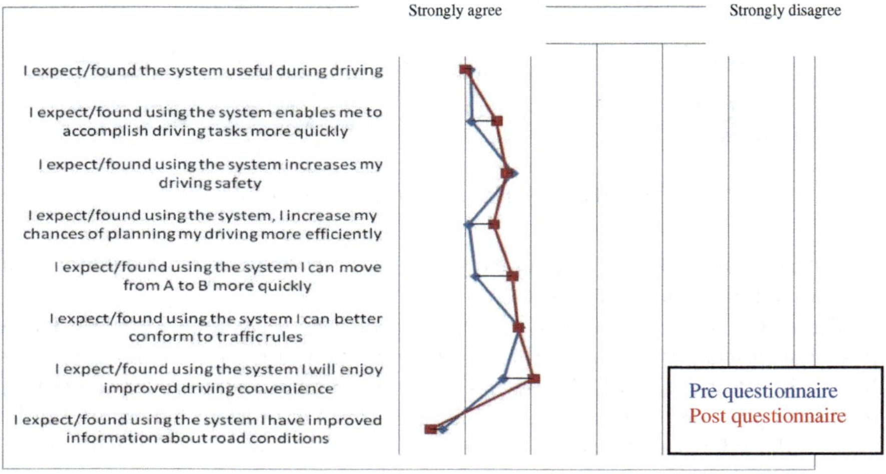

Fig. 3 Results for "Perceived Usefulness" of the COOPERS system

Figure 4 shows the results of the pre- (in green) and after- (in pink) questionnaire concerning the construct "Perceived Ease of Use". 'Ease of Use' is, along with 'Usefulness', the strongest indicator of Technology Acceptance. The figure shows that every single question asked in connection with how easy the system is perceived to be was already linked to quite high expectations of the users. On average, the test persons stated that they strongly agree that the interaction with the system was clear and understandable and that they find the system easy to use.

As described above, the test drivers' expectations of the system were outperformed by the system in terms of how easy the system is to use and how useful the system is for driving. The test drivers were also asked whether they intended to use a COOPERS system in the future. Figure 5 shows the results for the "Intention to Use" of the COOPERS system. Again, some of the answers given to the questions after experiencing the system were more positive than before experiencing the system. After using the system, on average, the respondents agreed that they would use a system if they had one, recommend the system to friends and buy the system when commercially available.

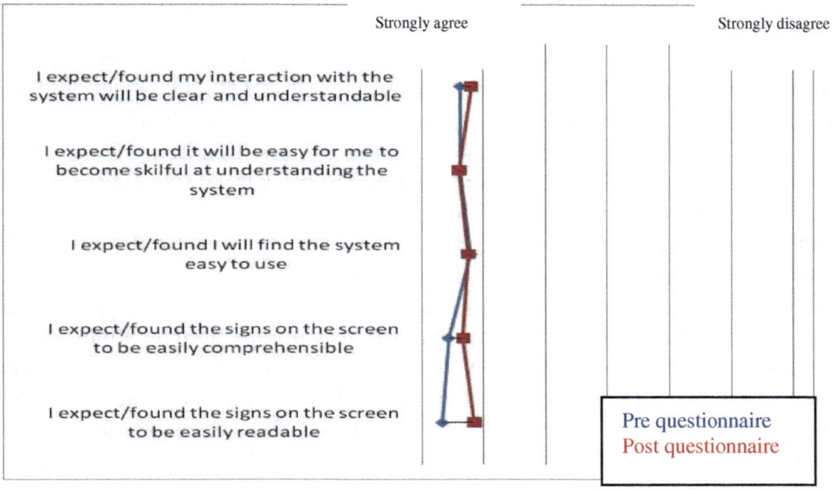

Fig. 4 Results for "Perceived Ease of Use" of the COOPERS system

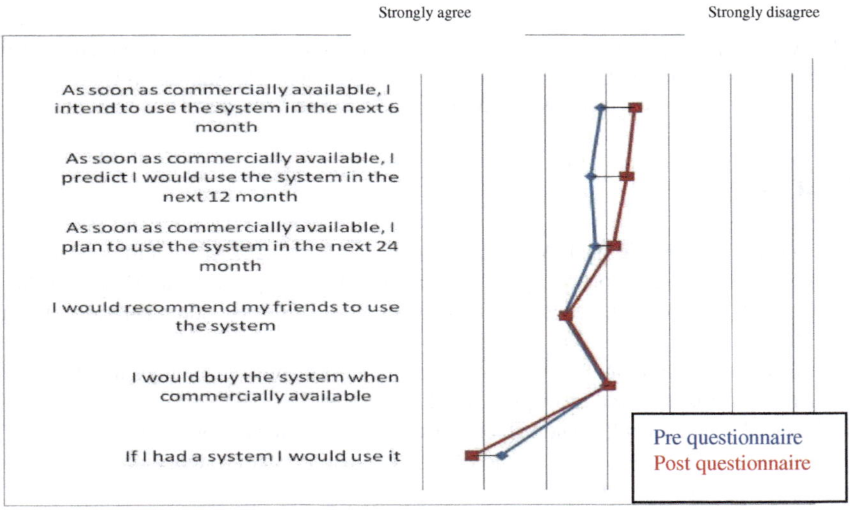

Fig. 5 Results for "Intention to Use" of the COOPERS system

Overall, the test drivers reacted in a very positive way to the COOPERS system. All major indicators for User Acceptance rank exceptionally highly, and are positive in a before/after comparison. The user interface and the useful COOPERS services seem to have made a good impression on the end users targeted. In general, the results achieved in the field test study are promising.

5 Conclusion and Limitations

User acceptance assessments allow for the provision of insight as to how new systems will be accepted by the general public as well as helping to gain an insight into the market opportunities for new technologies. User feedback is a valuable contribution for the development and redesign of user-friendly systems. In addition to this, next step estimations about the future market penetration are possible. High market penetration can be a precondition for the economic success of new technologies from one perspective but from another, it allows for the optimization of safety and efficiency impacts of telematic systems.

Regarding the results gained in the field test of the COOPERS project, it needs to be considered that these initial results give very promising indications as to potential user acceptance. Technology Acceptance measurement of the COOPERS system show that end user expectations are high as regards co-operative services. However, experiences in the field tests outperform these expectations considerably. Test persons said that they were keen to buy a COOPERS system as soon as they are commercially available.

Long-range adaptive behavior might reduce the benefits of some ITS services. This effect has not been investigated in the current study and needs further research in future methodological designs of ITS simulator and field tests.

Acknowledgements. The following individuals have contributed to the elaboration of the related reports within the COOPERS project: Susanne Fuchs, Alexander Frötscher, Martin Böhm, Christoph Hecht, Robert Kölbl, Meng Lu, Andy Richards, and Thomas Scheider.

References

1. Fuchs, S.: Organizational Adoption Models for Early ASP Technology Stages. In: Adoption and Diffusion of Application Service Providing (ASP) in Marketing Management, p. 251. University of Economics and Business Administration, Vienna (2005)
2. Morris, M.G.: A longitudinal examination of information technology acceptance, in School of Business. Indiana University, Indiana (1996)
3. Hong, W., et al.: Understanding continued information technology usage behaviour: A comparison of three models in the context of mobile internet. Decision Support Systems 42(3), 1819–1834 (2006)
4. Kamel, G.E.: The impact of technology acceptance consideration on system usage, and adopted level of technological sophistication: An empirical investigation. International Journal of Information Management 17(4), 249–260 (1997)
5. Venkatesh, V., et al.: User Acceptance of Information Technology: Toward a Unified View. MIS Quarterly 27(3), 425–478 (2003)
6. Fishbein, M., Ajzen, I.: Belief, Attitude, Intention and Behaviour: an introduction to Theory and Research. Addison-Wesley, MA (1975)
7. Davis, F.D.: Perceived usefulness, perceived ease of use, and user acceptance of information technology. MIS Quarterly 13(3), 319–340 (1989)

8. Venkatesh, V., Speier, C.: Computer Technology Training in the Workplace: A Longitudinal Investigation of the Effect of the Mood. Organizational Behaviour and Human Decision Processes 79(1), 1–28 (1999)
9. Davis, F.D., Bagozzi, R.P., et al.: Extrinsic and Intrinsic Motivation to Use Computers in the Workplace. Journal of Applied Social Psychology 22(14), 1111–1132 (1992)
10. Ajzen, I.: The Theory of Planned Behaviour. Organizational Behaviour and Human Decision Processes 50, 179–211 (1991)
11. Taylor, S., Todd, P.A.: Assessing IT Usage: The Role of Prior Experience. MIS Quarterly 19(2), 561–570 (1995)
12. Triandis, H.C.: Interpersonal Behaviour. Brooke/Cole, Monterey (1997)
13. Thompson, R.L., et al.: Personal Computing: Toward a Conceptual Model of Utilization. MIS Quarterly 15(1), 124–143 (1991)
14. Rogers, E.M.: The Diffusion of Innovations, 4th edn. Free Press, New York (1995)
15. Bandura, A.: Social Foundations of Thought and action: A Social Cognitive Theory. Prentice Hall, Englewood Cliffs (1986)
16. Anderson, J.E., Schwager, P.H., et al.: The Drivers for Acceptance of Tablet PCs by Faculty in a College of Business. Journal of Information Systems Education 17(4), 429–440 (2006)

Developments within the Scope of the German Test Site for Road Weather Stations

S. Piszczek, S. Grošanić, and A. Dinkel

Abstract. As the data for road weather stations is used for online traffic control within section control systems, it is very important for the efficiency of the traffic control systems to be based on reliable data of a high quality.

Therefore, a Test Site for checking the quality of road weather stations was established near Munich in Germany in 2003 and has been operational since then.

In close co-operation with all participants (sensor manufacturers, road authorities, German Federal Research Institute, research and consultancy bodies), the overall goal was to improve the sensors' quality as well as to establish methods to detect failures in measurements.

Furthermore, several improvements were carried out within the scope of the Test Site using the expertise of all participants and the infrastructure of the Test Site. The developments, reports and results obtained are both significant and helpful for manufacturers, road authorities, practitioners, research and consultancy.

1 Introduction to the German Test Site for Road Weather Stations

High quality data from road weather stations is essential for a fully functional traffic control system. In reality, the quality is mostly not monitored very well. Errors are often identified too late or by chance. In some cases, false weather detections are even not noticed.

S. Piszczek
Federal Highway Research Institute (BASt), Brüderstraße 53,
D-51427 Bergisch-Gladbach
e-mail: Piszczek@bast.de

S. Grošanić
Chair of Traffic Engineering and Control, Technische Universität München,
Arcisstraße 21, D-80333 München
e-mail: slavica.grosanic@vt.bv.tum.de

A. Dinkel
Transver GmbH, Maximilianstraße 45, D-80538 München
e-mail: dinkel@transver.de

J. Düh et al. (Eds.): Data and Mobility, AISC 81, pp. 49–60, 2010.
springerlink.com © Springer-Verlag Berlin Heidelberg 2010

Given the importance of the quality of the weather data detection, the German Test Site for Road Weather Stations was established near Munich Airport on the motorway A92. The Federal Ministry of Transport, Building and Urban Affairs (BMVBS) and the Federal Highway Research Institute (BASt) support the Test Site's finances and organization. The working group (AK) 3.2.1 of Germany's Road and Transportation Research Association (FGSV) operates as a supervisory board for the project. The South Bavarian Highway Authority (ABDS) is responsible for the operation of the Test Site while the monitoring and evaluation of the different sensor types is carried out by the Chair of Traffic Engineering and Control at the Technische Universität München. The following figure shows the German Test Site for Road Weather Stations on motorway A92.

Fig. 1 German Test Site for Road Weather Stations

At the Test Site, three Webcams are installed. The pictures are updated every minute and are available at http://www.vt.bv.tum.de/umfelddaten.

The sensor systems are evaluated in terms of their applicability for traffic control. Furthermore, the Test Site is used for enhancing the quality of the hardware and software of the sensor systems. With the knowledge gained from the evaluated data from the Test Site, the specifications for tender and the demand and operation of road weather stations are established. Each year, the German federal departments for road construction are informed about the results of the evaluated sensor types from the Test Site.

The following two tables show the different measurements and the number of sensors situated at the Test Site. The measurements are categorized into primary and secondary types of measurements. The primary measurements (**Table 1**) are used directly for traffic control purposes.

Table 1 Number of primary measurements at the Test Site (2010)

Primary measurement	Number of sensors
Precipitation intensity	8
Precipitation type	6
Status of the road surface	4
Visibility	10
Water film thickness	4

The secondary measurements (**Table 2**) are used for plausibility checks. These measurements are used for some special applications like warnings at bridges with heavy gusts of wind.

Table 2 Number of secondary measurements at the Test Site (2010)

Secondary measurement	Number of sensors
Air temperature	4
Average wind speed	4
Direction of wind	4
Freezing temperature	3
Ground temperature in depth 1	1
Maximum wind speed	4
Melting point temperature	2
Relatively air humidity	4
Road surface temperature	4

All of the weather measurements are detected in one minute intervals [1]. With special databases configured for the Test Site, the measured data can be analyzed for definite time periods. The weather data is then analyzed in terms of

- accuracy,
- reaction time at the beginning of a weather occurrence and
- intensity of the weather situation.

At this Test Site, different suppliers of road weather stations were checked under the same conditions. Manufacturers which are not directly from the typical sensor market for Road Weather Stations also took part with prototypes, e.g. metrological or automotive suppliers. The operation of the Test Site is made possible by constructive team work based on the following manufacturers which are currently involved at the Test Site:

- Adolf Thies GmbH & Co. KG, Göttingen
- Boschung Mecatronic AG, Granges-Paccot, CH
- G. Lufft Mess- und Regeltechnik GmbH, Fellbach
- Ott Messtechnik GmbH & Co. KG, Kempten
- Sick Maihak GmbH, Analysen- und Prozessmesstechnik, Reute
- Ingenieurbüro Spies GbR, Hohenwart
- Vaisala Traffic GmbH, Hamburg

With the supply of sensors to the Test Site, the manufacturers have the possibility of comparing their own systems with other competing systems or with reference measurements. Through the permanent monitoring of the data from the sensors, the manufacturers can update and enhance their sensor systems.

2 Results

The following section describes several works which were carried out within the German Test Site for Road Weather Stations. The sensor systems are evaluated concerning their suitability for traffic control systems. A huge amount of data available from 2003 onwards was used to develop tools for evaluating the data quality. These offline tools are used to help detect erroneous measurements and analyze the long-term behavior of data from road weather stations. The developed tools can also be used for checking "standard" road weather stations.

Furthermore, the relationship between weather influences on traffic behavior is analyzed. For this research, a mySQL-database was developed which allowed an easy comparison of weather and road traffic data. Several analyses were carried out in order to quantify the effects of weather on road traffic safety.

2.1 Annual Reports

The annual reports from the Test Site contain the evaluation results for the different sensor types. The results are based on data-driven evaluation, involving the data from one year. The evaluation has been carried out in this way since 2005.

The overall assessment is based on the results of the sensor data concerning the plausibility of the results and the reaction time. The plausibility of the results is determined:

- relative to other results (continuously),
- thoroughly based on observation (random samples) and
- thoroughly based on a reference (random samples).

In coordination with the manufacturers, reference methods have been established to evaluate several parameters (e.g. a spraying box for checking measurements of water film thickness). Furthermore, a qualitative comparison of time-variation curves is established. The measured data is compared to reference measurements (e.g. sensor pluviometer for precipitation intensity, psychrometer for temperature) and reference observation (analysis of webcam pictures for visibility). Any differences identified are analyzed accurately. Furthermore data from the different measuring methods was correlated and analyzed. These analyses are assisted by field monitoring.

In Table 3, the valuation method for every reviewed primary measurement is shown.

Table 3 Overview of the valuation method for primary measurement

Primary measurement	Valuation method
Precipitation intensity	Pluviometer, time-variation curve
Precipitation type	Webcam, field monitoring
Visibility	Webcam, time-variation curve, field monitoring
Status of the road surface	Webcam, field monitoring
Water film thickness	Time-variation curve, field monitoring, spraying box, towel test

For the secondary measurements, the time-variation curves were compared qualitatively.

If more criteria are analyzed, the results of several valuation methods are averaged. In terms of the applicability of the sensors for detecting critical weather situations for traffic control, an evaluation scheme was developed and applied. The sensors were classified as:

- "suitable": the reviewed sensor responds to situations promptly and represents them well. The sensor can be used in traffic control systems.
- "suitable with restrictions": the reviewed sensor responds to situations and represents them but has delays in reaction time or partially incorrect amplitudes. The reasons for an unsuitable application can be calibrated in most cases.
- "unsuitable": the reviewed sensor does not respond to situations promptly and represents them insufficiently.

The evaluations for the measurement of water film thickness are based on a test using a 'spraying box'. A known water film thickness is applied to the sensor with the help of a computer-controlled airbrush spray gun that applies specified amounts of water to the sensor surface [2]. During the period 2005-2006, insufficient tests were carried out using the spraying box, so the basis of the evaluations was too small and no classification was made.

The following figure shows the annual results of the sensors for the primary measurements taken during the testing phases from 2005 until 2008. One evaluation phase spans from November until October of the following year.

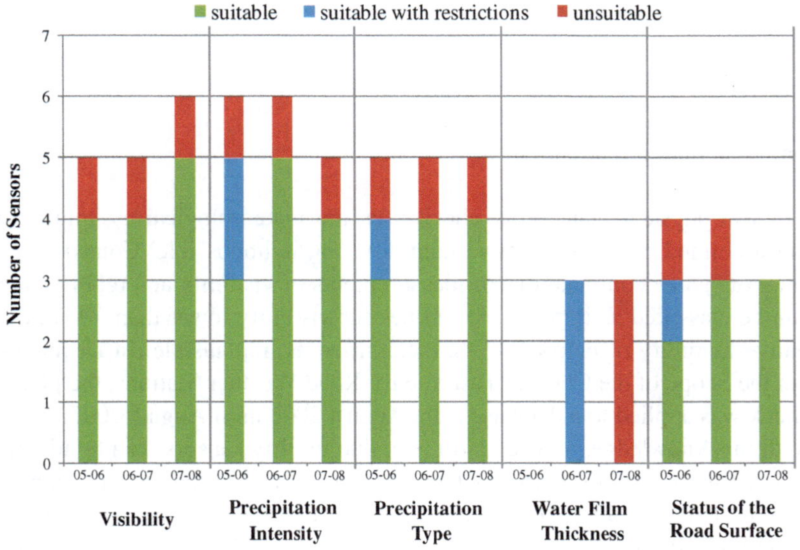

Fig. 2 Evaluation of the sensors for primary measurements [3], [4], [5]

In Development within the Scope of the German Test Site for Road Weather Stations, the results of the evaluation during the three testing phases are shown. In the three evaluation phases, most of the sensors for visibility were evaluated as suitable for use in traffic control systems. Only one sensor is classified as "unsuitable" for use in traffic control. During the evaluation phase from October 2005 until November 2006, the measurement of precipitation intensity was evaluated as "suitable for use in traffic control" for three sensor types and two sensor types were evaluated as "suitable with restrictions". During the following testing phases, no sensor was evaluated as "suitable with restrictions". The evaluation results for precipitation type and status of the road surface were also classified in a similar way.

The measurement of water film thickness is evaluated as "unsuitable" (in testing phase 07-08) or "suitable with restrictions" (in testing phase 06-07). This result shows that, for this measurement, there is still potential for optimizing the sensor systems.

2.2 Benchmarking and Data Distribution Tool

Validation and quality management is a major task of data processing to detect any malfunctioning data sources and exclude them from further usage in traffic control. By now, a huge database has been developed which enables analysis of the correlation between the measured values based on data mining methods to perform benchmark tests and error detection.

A prototype software tool has been implemented to apply plausibility checks to the database. The following plausibility checks have been developed, tested and optimized:

- plausibility checks for single measurements,
- checking logical-physical coherence and
- long term plausibility checks.

The logical-physical plausibility check uses the meteorological assumption that during a defined period of relative humidity, fog is impossible. Considering this assumption, some combinations of measurements of visibility and relative humidity can be classified as implausible. Values for visibility lower than 500 m during a relative humidity of below 90 % are classified as implausible. In Development within the Scope of the German Test Site for Road Weather Stations, the plausibility check was applied to a database from August 2006 until August 2007.

With this knowledge, a false alarm (e.g. due to dirty sensors) can be identified. A data distribution tool is currently being developed and tested, which can be used directly in daily operation. The weather data is analyzed automatically for plausibility and long-term behavior. The results are visualized at a number of different levels of slickness and moisture. The data distribution tool is configured so that it is portable and can be used at every road weather station.

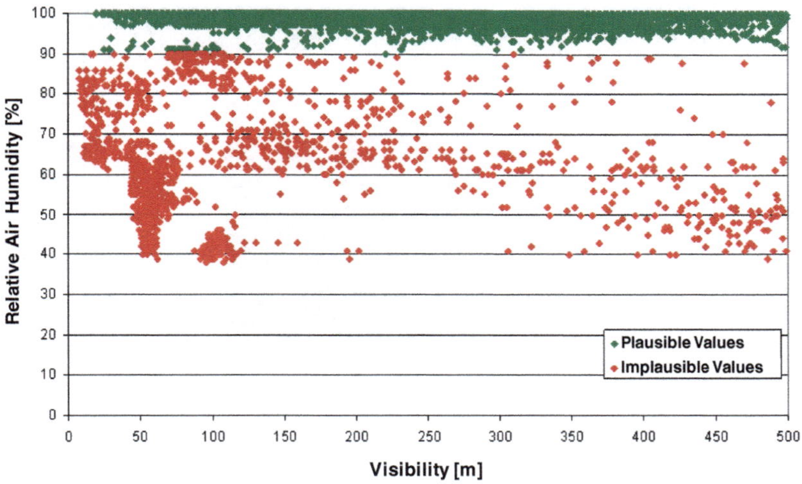

Fig. 3 Result of the logical-physical plausibility check for visibility-measurements [6]

Detection and usage of measurements from road weather stations as well as these plausibility checks will be published as a technical bulletin [7]. This technical bulletin will help in the acquisition and usage of road weather data in German traffic control systems.

2.3 *Visibility Tool*

Over the years, several hundreds of thousands of webcam pictures (one per minute) were collected and stored for evaluation. To save time in searching for situations with reduced visibilities, a rough visibility estimation can be carried out initially using the webcam and appropriate automatic image processing algorithms (pre-processing of the manual observed visibility estimation). An algorithm has been developed [8] and tested [9]. It consists of the following steps:

For each image:

1. Calculation of grey values (matrix of grey values)
2. Luminance estimation (image usable: yes/no)

If image is usable:

1. Convolution with edge detector (edge-matrix)
2. Visibility estimation (edge intensity (sum, medium, maximum) → visibility in “pixel-rows” → visibility in meters

The calculated visibility results are compared to the individual evaluation process. A trained person analyzes the webcam images and checks the visibility distance. For each minute, the defined objects that were or were not visible was recorded. The defined distances of the objects at the Test Site are shown in Development within the Scope of the German Test Site for Road Weather Stations.

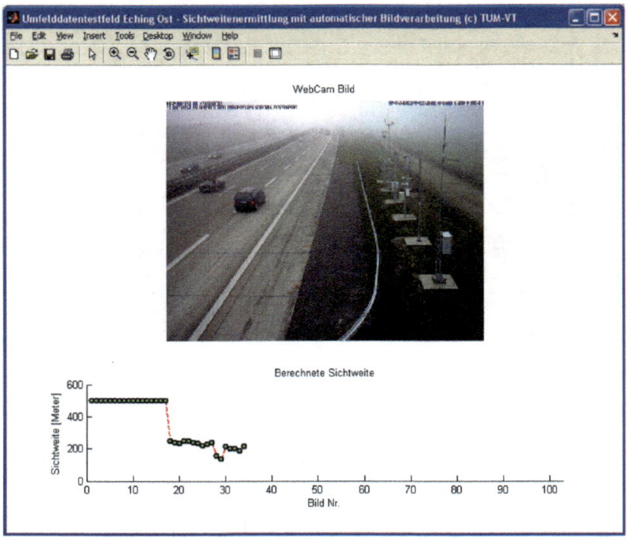

Fig. 4 Graphical output image and calculated visibility [8]

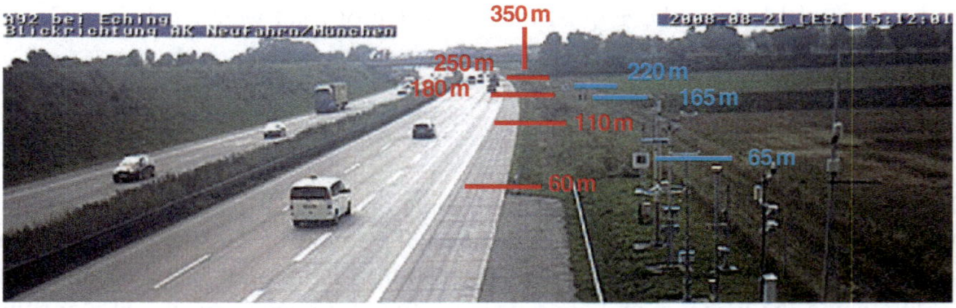

Fig. 5 Defined distance of objects for visibility [5]

The value for visibility was defined as the difference between the distance of the object that was still visible (minimal visibility) and the distance of the object that is no longer visible (maximum visibility), the so called "visibility corridor".

Uncertainties due to the subjective perception of the human observer, differences between meteorological and perceived visibility and position during observation were resolved in the sensor's favor.

The following picture shows the visibility corridor (black lines) and the calculated visibility from the visibility tool (blue line). The grey lines are the visibilities of the different sensors at the Test Site.

This graph shows that the visibility tool can be used for a rough estimation of situations with reduced visibility.

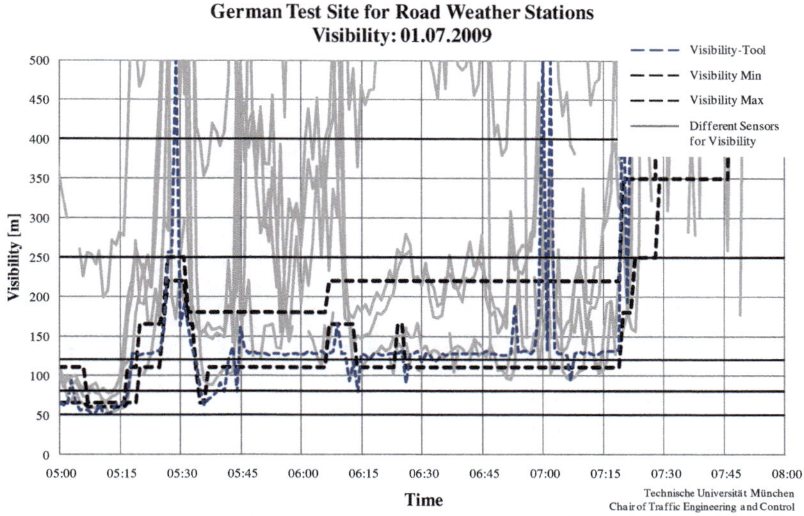

Fig. 6 Visibility corridor and calculated visibility

2.4 *Effects of Weather on Road Traffic Safety*

High traffic loads and traffic congestion lead to increased risks of traffic accidents and thus to a reduction in traffic safety. An important objective of traffic control systems is the increase in traffic safety through a dynamic reaction to current traffic and road weather and road surface conditions. Detecting road weather and road surface conditions and deriving corresponding suggestions for the variable message signs (warning signals, speed limits) are thus an important feature of section control systems. Traffic control systems use variable message signs (VMS) which have an effect on traffic safety and the relevant traffic flow parameters (e.g. speed, special attention in the case of potential risk situations).

To better understand the coherence between critical weather situations and traffic safety, some relevant indicators have been analyzed. In the first step, it had to be determined which safety parameters (e.g. velocity, time-to-collision, net time gap) are useful for determining traffic safety during different weather situations. Therefore a study was undertaken [10] to identify which safety indicator is best for each situation such as a reduction of visibility or precipitation. The safety indicators employed consider situations before a crash happens. The result of analyzing a database comprising half a year's worth of data was that reduction of visibility has an effect on the net time gap and the Compensated Platoon Braking Time Risk (CIBTR), which is described in [11]. Rainfall influences the traffic flow in terms of reduced velocities and the variance of velocity.

In Development within the Scope of the German Test Site for Road Weather Stations, the velocity (green line) and the net time gap (light blue line) are shown for intense rain (max. precipitation about 20 mm/h, red line). The data is taken

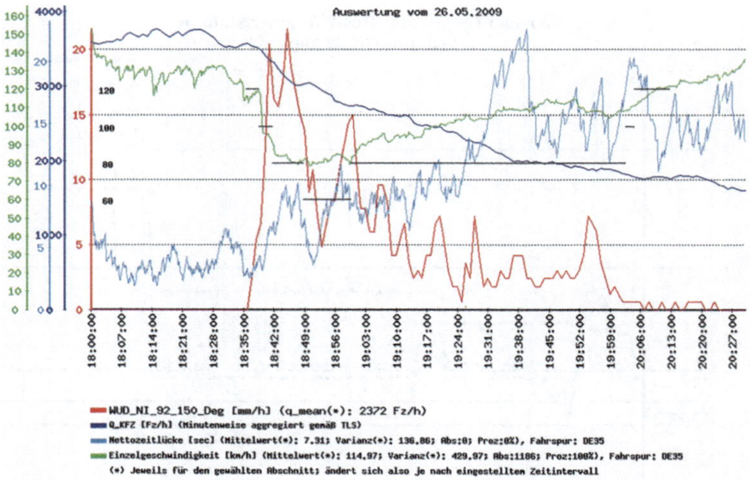

Fig. 7 Effects of precipitation on traffic behavior [10]

from detectors near the Test Site for Road Weather Stations on motorway A92 (towards Munich). The traffic volume (dark blue line) is about 2400 veh/h.

In the time period before the rainfall started, the net time gap between consecutive vehicles was 4 seconds; during the precipitation, the net time gap rose to 10 seconds. An influence on traffic flow can therefore be deduced. The number of small time gaps is reduced during the rainfall from 27% to 9%. The variance of velocity is lower during precipitation situations compared to during dry road surface conditions.

Furthermore, different precipitation situations were analyzed. They were grouped into 21 categories based on duration and intensity of precipitation. The sensitivity of the safety indicators for velocity and variance concerning a change in traffic parameters was proven [10].

The results can be used for further studies on this topic.

2.5 Offline Analysis with mySQL-Database

A mySQL-database, usable offline, was developed [10]. With this database it is possible to integrate different archive files and different data types together in one system. At the intersection on motorway A92 near the Test Site, the traffic data is also collected as single vehicle data by the South Bavarian Highway Authority. The data is collected when each vehicle passes the detection field, generating a new timestamp and recording the velocity of this single vehicle in the database. Furthermore, the data from road weather detectors, loop detectors and VMS-status is available for every minute [1].

To connect these two different resources in one single database, a mySQL-database was developed. With this mySQL-database, offline analyses and visualization of weather and road traffic data can be performed. It is possible to visualize different safety parameters like time to collision or (compensated) individual braking time risk together with velocity and traffic flow. The evaluation of the relationship between traffic flow and different weather situations is possible. This tool is being used for further research on the relationship between traffic flow and different weather situations.

3 Conclusion

Within the scope of the Test Site, various works have been carried out. The Test Site project has shown that not all of the tested sensors are able to meet expectations in terms of sensor quality. In daily operation, road weather data will not be analyzed as intensively as at the Test Site project. The developed plausibility checks will help to improve data quality. All of the important aspects for the acquisition and usage of road weather data in German traffic control systems are summarized in a technical bulletin [7]. This bulletin will be published soon.

The measurements of the road weather stations are used directly as input data for VMS, so high data quality is very important. An authentic variable traffic sign is the basis for a high acceptance level on the part of the road user. The correlation between special weather conditions and the changes in traffic flow have to be analyzed further. The general goal is to improve the acceptance of traffic control systems.

References

1. Technische Lieferbedingung für Streckenstationen (TLS), issued by Bundesministerium für Verkehr, Bau und Stadtentwicklung (2002)
2. Badelt, H., Breitenstein, J., Scheuerl, D., Häusler, G.: Referenzmessverfahren für die Prüfung von Niederschlagssensoren, Abschlussbericht Projekt 01670 der Bundesanstalt für Straßenwesen, Bergisch Gladbach (2003)
3. Dinkel, A., Leonhardt, A., Busch, F.: Umfelddatenerfassung in Streckenbeeinflussungsanlagen, Testfeld "Eching Ost" des Bundes, Abschlussbericht 2. Testphase, issued by Bundesministerium für Verkehr, Bau und Stadtentwicklung (2007)
4. Dinkel, A., Leonhardt, A., Busch, F.: Umfelddatenerfassung in Streckenbeeinflussungsanlagen, Testfeld "Eching Ost" des Bundes, Abschlussbericht 3. Testphase, issued by Bundesministerium für Verkehr, Bau und Stadtentwicklung (2008)
5. Dinkel, A., Grošanić, S., Busch, F.: Umfelddatenerfassung in Streckenbeeinflussungsanlagen, Testfeld "Eching Ost" des Bundes, Abschlussbericht 4. Testphase, issued by Bundesministerium für Verkehr, Bau und Stadtentwicklung (2009)
6. Dinkel, A., Leonhardt, A., Busch, F.: Umfelddatenerfassung in Streckenbeeinflussungsanlagen, Testfeld "Eching Ost" des Bundes, Dokumentation des Benchmarking-Tools zur Überprüfung von FG3-Messwerten, internal document (2008)
7. FGSV AK 3.2.1: Hinweise zur Erfassung- und Nutzung von Umfelddaten in Streckenbeeinflussungsanlagen, Wissenspapier, draft version (2010)
8. Leonhardt, A., Dinkel, A., Busch, F.: Umfelddatenerfassung in Streckenbeeinflussungsanlagen, Testfeld "Eching Ost" des Bundes, Automatische Auswertung von WebCam Bildern zur Sichtweitenermittlung, internal document (2008)

9. Ungureanu, T.: Kalibrierung und Validierung des Sichtweiten-Tools zur automatischen Auswertung von WebCam-Bildern zur Sichtweitenermittlung, Studienarbeit, Munich (2010)
10. Maget, C.: Untersuchung des Wettereinflusses auf den Verkehr, Diplomarbeit (2010)
11. De Mouzon, O., El Faouzi, N.-E., Pham, M.-H., Chung, E.: An Effective Real-Time Proactive Road Traffic Risk Indicator Based on Traffic Data: Compensated Platoon Braking Time Risk (CPBTR), Geneva (2008)

Documentation of Flood Damage on Railway Infrastructure

Andrew P. Moran, Annegret H. Thieken,
Andreas Schöbel, and Christian Rachoy

Abstract. Floods pose a considerable risk to the infrastructure and the operation of the Austrian Federal Railway network (ÖBB) as seen by the latest major flood event at the March River in 2006. This event, which led, amongst other things, to the shutdown of the Austrian Northern Railway Line between Gänserndorf and Břeclav (Czech Republic) for several months and caused direct infrastructural damage totaling several million EURO, revealed once again the significant vulnerability of the extensive rail network to natural hazard processes in Austria. Experiences gained during this and other events inspired the development of standardized and pragmatic documentation of flood damage. This is aimed in particular at aiding those who inspect the damage during or in the early stages of the aftermath and serves as a guideline for simplifying the damage assessment process, helping to estimate the extent of the losses and producing consistent information. In the longer term, consistent documentation of flood losses can support the development of loss estimation models and efficient risk management strategies. According to this proposal, the documentation takes place in four categories: event documentation, damage documentation, object information and information concerning damage reduction. Thereby, important object categories are defined such as the railway cross section, bridges, interlocking blocks, (station) buildings and transformer substations, to which each of the five damage classes are attributed. These classes represent both structural damage as well as operational impacts. Through visualization in a GIS (Geographical Information System) and by linking information on structural damage to hydraulic data, further insight into the damage processes is gained. The documentation is applied for the 2006 floods at the March River.

Andrew P. Moran . Annegret H. Thieken
alpS – Centre for Natural Hazard and Risk Management,
Grabenweg 3, A-6020 Innsbruck
e-mail: moran@alps-gmbh.com

Andreas Schöbel
Institut für Verkehrswissenschaften, TU Wien, Karlsplatz 13/230-2, A-1040 Vienna

Christian Rachoy
ÖBB-Infrastruktur AG, Nordbahnstr. 50, A-1020 Vienna

J. Düh et al. (Eds.): Data and Mobility, AISC 81, pp. 61–70, 2010.
springerlink.com © Springer-Verlag Berlin Heidelberg 2010

1 Introduction

Risks induced by various natural hazard processes are significant in the Alpine country of Austria. Of the different types of natural phenomena that can have an effect on the widespread rail network floods, as discussed in Nester et al. (2008), pose a great threat and clearly cause a substantial amount of damage to rail infrastructure. Thus, floods induce some of the highest monetary losses for the Austrian Federal Railway ÖBB. However in spite of the great relevance of the topic, there seems to be little loss data and few models available both within Austria as well as internationally. The development of damage estimation models often takes place on an empirical basis. Therefore, this study aims to improve data availability in the long term by developing a standardized documentation scheme.

The concept of consistent flood damage documentation as part of a modern risk management strategy for railway infrastructure is elaborated and tested for the Austrian Northern Railway. It is subject to potential flooding, as recently observed during the widespread March River floods in April 2006.

This approach contains four main procedural measures. Firstly, the rail infrastructure data is closely examined and the most important infrastructural elements with regard to assets and operational relevance are identified. Subsequently, these are allocated to various object classes of which each of these object classes is assigned five main damage classes. This classification contains both direct damage as well as indirect impacts. In an example application of this procedure, data from the 2006 flood event and flood simulation models are applied in order to determine the affected assets. Finally, a standardized catalogue of the documentation of damages is developed. The paper closes with some remarks on the implementation of the proposed procedures.

2 Data and Methods

2.1 The Austrian Northern Railway Line

The Austrian Northern Railway line of the Austrian Federal Railway (ÖBB) was selected for the development and testing of the methods discussed in this paper. This railway line has at least two tracks, is electrified across its entire length and belongs to the ÖBB's core rail network. It was chosen because of its length (approx. 80 km from Vienna to Břeclav), making it a suitable test stretch for the issues in hand, and due to the flood risk stemming from the March River and its tributaries. Named Kaiser-Ferdinand-Nordbahn, it was opened in 1837 to connect Vienna to the coal regions of Silesia (Mayerhofer 2006). During the Cold War, the proximity to Czechoslovakia led to a severe loss in its significance. However, since the lifting of the Iron Curtain in 1989, the importance of the Northern Railway Line has steadily increased again over the last two decades (Mayerhofer 2006). It now plays a vital role in connecting Eastern Austria (Vienna) with the Czech Republic (Prague, Ostrava) and Poland (Kraków, Warsaw).

2.2 *March River Flood Event 2006*

The March River flood event in April 2006 was the result of extraordinary deep and widespread snow cover throughout the entire March River watershed followed by the rapid and intense onset of snowmelt in combination with significant precipitation. While the snowfall level rose abruptly to 2000 m a.s.l., a precipitation amount of 30 mm was measured at the Haugschlag gauge on March 25 and 26. It was followed by another 30 mm on March 28 and reached up to 50 mm in the Czech part of the watershed area (BMLFUW 2006). The March River began to flood its banks on March 28 and finally the dams were breached on April 4 and 5 in the area of the settlements of Jedenspeigen, Sillfried and Grub (BMLFUW 2006).

The recent occurrence of this large scale flood event shed light on the vulnerability of such an infrastructure network to natural hazard processes. The flood event caused substantial damage to the railway along a section of around 10 km, leading to the entire shutdown of all passenger and freight operations for months, making replacement transport services necessary.

2.3 *Data*

The data available contains information about the railway infrastructure and the March River flood event. A number of photos taken during and directly after the flood enabled the event to be reconstructed and helped the corresponding damage to be recorded. Spatially referenced geodata of the Austrian Northern Railway in combination with flood model results, e.g. the Austrian-wide flood model HORA with recurrence intervals of 30, 100 and 200 years (Willems 2006), formed the basis for the identification of risk-prone railway sections for different flood scenarios. Furthermore, book keeping data containing information on the financial assets of the railway's infrastructure was made available.

3 Results and Discussion

3.1 *Classification of Structural Damage for Different Important Railway Infrastructure Elements*

An analysis of the Austrian Northern Railway's assets, in combination with an investigation of the operational relevance of the various objects, led to the definition of five main object classes: railway cross sections, bridges, interlocking blocks, (station) buildings and transformer substations. Objects belonging to these classes are characterized by high asset values and play an important part in the performance of rail operations.

Each object class was allocated a further five structural damage classes that describe the physical structural damage as well as potential indirect effects with operational relevance. These classes form the basis for flood damage documentation and are depicted as an example for railway cross sections and bridges in Table 1, Table 2 and Figure 1. In future, these classes will also be used to derive damage estimation models for each object class.

Table 1 Description of structural damage to railway lines/cross sections and possible indirect effects

#	Damage grade	Description of direct structural damage	Description of possible indirect effects
1	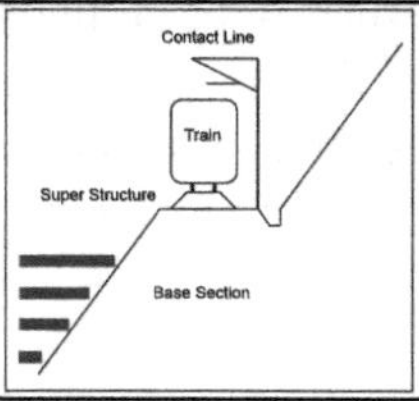	Flooding reaches the base section without any notable damage.	none
2	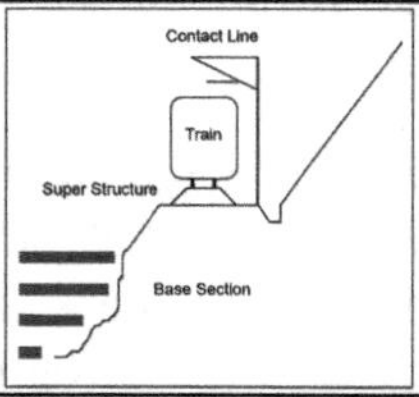	Flooding reaches the base section. Erosion occurs.	Possible operational limitations (slower speeds, delays).
3	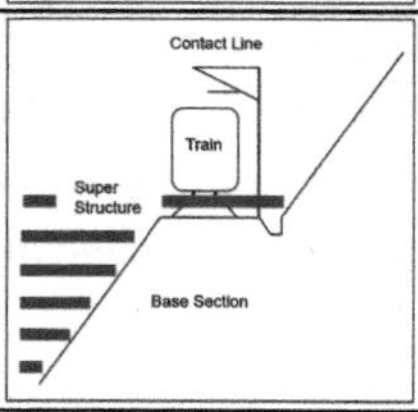	Track super structure is flooded.	Track is closed. Detours or replacement transport services are necessary.
4	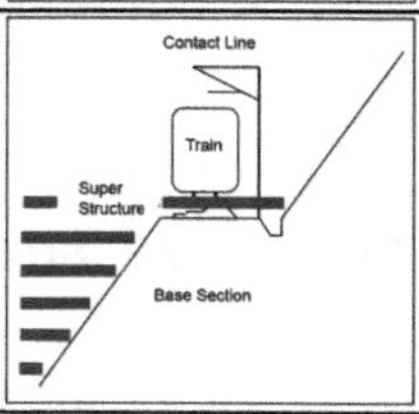	Erosion of the track super structure. Complete reconstruction is necessary.	Detours or replacement transport services for several days are necessary.
5	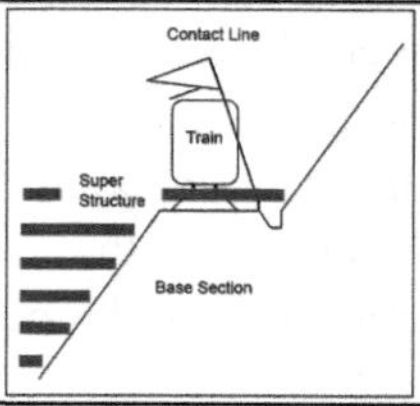	Additionally the overhead contact line, signals etc. are damaged.	Detours or replacement transport services for several weeks are necessary.

Table 2 Description of structural damage to railway bridges and possible indirect effects

#	Damage grade	Description of direct structural damage	Description of possible indirect effects
1	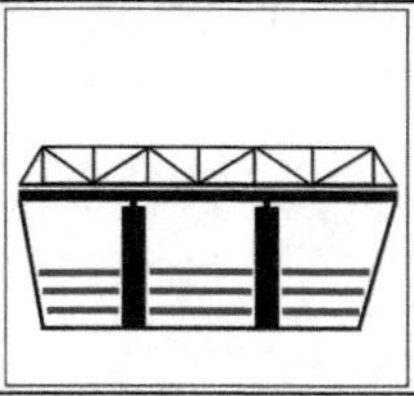	No adverse effects due to flooding.	none
2	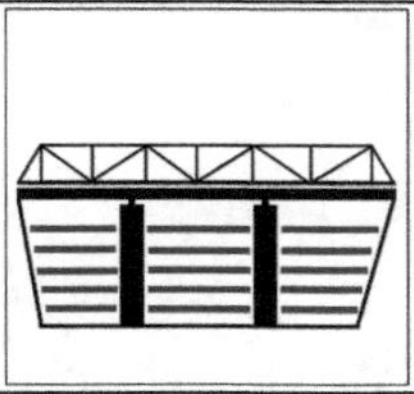	No restrictions, maintenance necessary in the course of the next inspection.	none
3	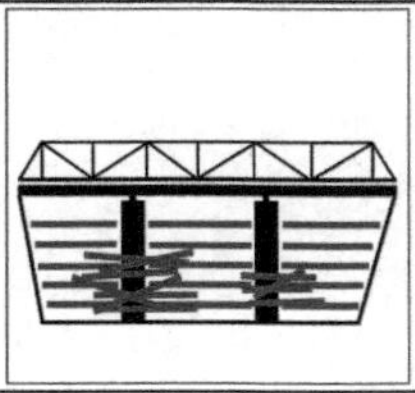	Log jam, removal is required.	In general no operational limitations. Speed reductions or short closures are possible during the removal of the log jam.
4	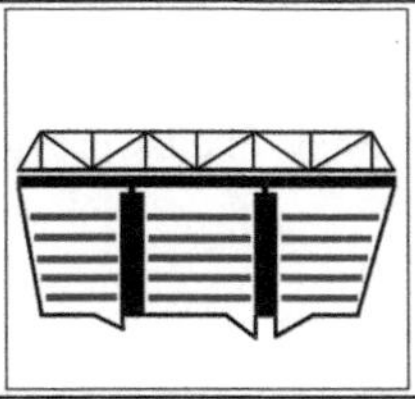	Formation of a scour. Repair is necessary.	Rail operations are restricted. Detours or replacement transport services may be necessary.
5		Scour or major structural damage to the object. Bearing strength his no longer guaranteed. Bridge is completely closed.	All rail operations are closed for many days or weeks.

3.2 Application of the Classification Scheme: Flood at the March River in 2006

In order to describe the damage that occurred during the flood event in 2006 along the Austrian Northern Railway line, the classification scheme for the standard railway cross section, as depicted in Table 1, was applied. Photographs taken during and in the direct aftermath of the flood event were assessed (Figure 1). Thereby damage grades of the standard cross section were assigned to the various

Fig. 1 Field examples of damage grades 1 to 4 to railway lines caused by flooding of the Austrian Northern Railway line by the March River in April 2006. Photos provided by ÖBB-Infrastruktur AG.

photos. Several photographs contain more than one class or sections at the transition between two classes. The kilometer measurement system along the railway allowed most flood damage to be spatially referenced. The results of the spatial allocation of damage classes are illustrated in Figure 2 in which the railway line is broken down into points at intervals of 100 meters respectively. Points to which no photographs could be allocated remain colorless.

Due to the spatial referencing of these structural damage classes, a correlation analysis with water height and flow velocity data can be performed, if detailed 2D hydraulic simulations become available. Thus, this forms the foundation for a damage model.

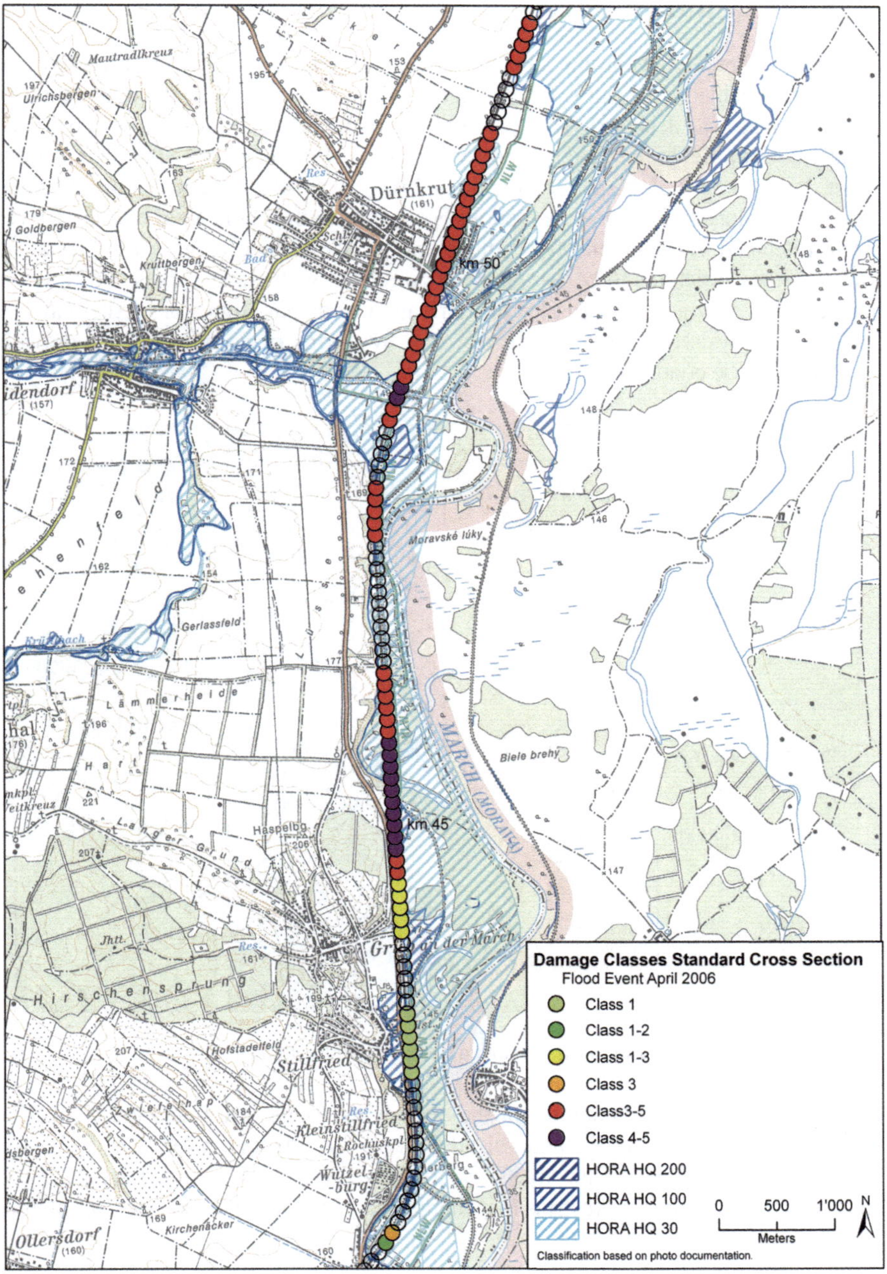

Fig. 2 Allocation of damage classes of the standard cross section between Angern and Dürnkrut for the 2006 March River flood event

4 Criteria for a Standardized Documentation of Damage to Railway Infrastructure

In order to improve the current data pool on flood damage to railway infrastructure and to derive reliable information for damage models, a standardized documentation of flood damage was designed. This proposal entails a number of criteria, which should ideally be acquired in the field. In cases where field work may not be possible, various other data sources can be accessed and combined to ascertain the necessary information for the standardized documentation. For example, the application of data derived from a detailed 2D hydraulic flood model would allow flood damage to objects and various other structures to be retroactively correlated with water depths and flow velocities, thus serving as a basis for the derivation of damage functions. Alternatively, an initial analysis of interdependencies (e.g. by means of correlation or regression analyses) can be conducted.

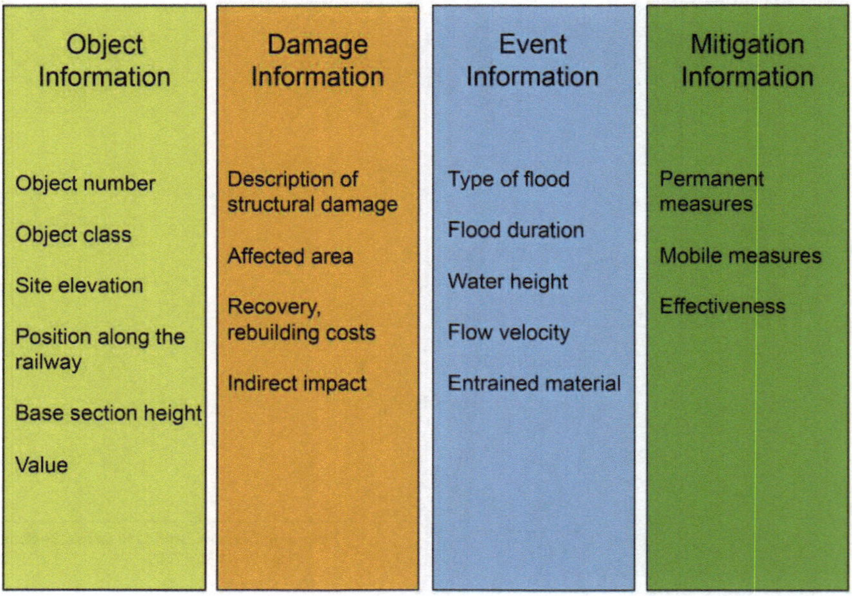

Fig. 3 Standardized damage documentation

The composition and contents of the criteria catalogues are based on Thieken et al. (2010), who developed a standardized damage documentation for various sectors. The criteria catalogue for the traffic sector was revised and adapted especially to railway infrastructure in Austria. In general, the criteria catalogue consists of four main components:

- Information regarding the damaged object (this information should be derived from data already in existence or should be linked to existing data sets of the railway company. The requirement for this would be a consistent and unambiguous object characterization (ID).)

- Information describing the damage
- Information about the event characteristics (placing the damaged object in a local context)
- Information regarding measures for damage mitigation.
- In Figure 3 the most important variables in each of the four categories are listed.

5 Towards Implementation

In order for a rapid and effective reaction to take place in the case of an event, appropriate types of organization, standardized documentation forms, responsibilities, etc. must be thought through, prepared for and implemented as much as possible, in plenty of time before the occurrence of an event. Thus, organizing the data acquisition is an integral part of these precautionary procedures. It is to be sufficiently adapted to the existing administrative and organizational framework.

The following aspects need to be taken into consideration in order to develop an organizational structure for the ascertainment of damage (Hübl et al. 2006):

- Definition of the goals and limitations for recording damage.
- Organization of data compilation and nomination of the person(s) responsible, (e.g. coupling data acquisition to the invoice for the repair costs or alternatively the approval of funds).
- Development and supply of materials and instruments (e.g. conveyance of the abovementioned criteria into documentation forms and maps. Consideration of whether GPS-handheld receivers and other instruments may be of importance.)
- Securing the linkage to external information (e.g. meteorological and book keeping data, land survey cadastre information, data synchronization and exchange between different departments).
- Provision of training courses and workshops.
- Conceptualization and configuration of a damage database and connecting it to the Geographical Information System of the Austrian Federal Railway (ÖBB) already in existence.
- Organization and investigation of the plausibility of the data input.
- Regulation of the data provision and data transfer.

Hübl et al. (2006) advise an adjustment in the amount of time and expenses needed to acquire data according to the type and significance of the event. Thereby, events are to be classified according to the size of the affected area, recurrence intervals and impact.

In order to ensure a successful and sustainable event documentation, implementation procedures should be developed by a working group comprising members of all the relevant departments of the enterprise as well as external advisors.

6 Conclusion

In the scope of flood risk analysis, a concept for standardized and consistent damage documentation has been successfully developed and applied to the railway infrastructure.

Further future issues to be dealt with entail the collection and processing of data from previous flood events and, if possible, the linkage of different data sources. In addition, a detailed data analysis is required in order to identify cause-effect-relationships between damage classes and event characteristics and to better understand damage processes. Finally, the damage documentation developed in the course of this study should be implemented within the corporation as a guideline for simplifying the damage assessment. This includes sheets for field work as well as organizational issues, database development and maintenance. The consistent documentation of damages can be a vital aid in the further development of loss estimation models and should be integrated into efficient risk management strategies.

Acknowledgements. This work was funded as part of the alpS-project 1.19ABC within the framework of the Austrian funding program Kplus of the Austrian Research Promotion Agency (FFG).

References

BMLFUW, Das Hochwasser an der March im Frühjahr 2006 – Beschreibung der hydrologischen Situation der Abteilung VII 3 (2006)

Hübl, J., Kienholz, H., Loipersberger, A. (eds.): DOMODIS – Dokumentation alpiner Naturereignisse. Interpraevent, Schriftenreihe 1, Klagenfurt (2006)

Mayerhofer, D.: Infrastrukturkonzept für den Ausbau der Nordbahn Wien-Süßenbrunn-Bernhardsthal. Diploma Thesis. Institute for Transport Studies, University of Natural Resources and Applied Life Sciences Vienna (2006)

Nester, T., Schöbel, A., Drabek, U., Kirnbauer, R., Rachoy, C.: Flood Warning System for the Austrian Railways. In: 1st Lakeside Conference on Safety in Mobility, Intelligent Weather Information Systems and Services in Traffic and Transport (2008)

Thieken, A., Seifert, I., Merz, B. (eds.): Hochwasserschäden – Erfassung, Abschätzung und Vermeidung, p. 304. Oekom-Verlag, Munich (2010)

Willems, W.: HORA. Teilprojekt A: Hydraulische Arbeiten (HORA I). Teilprojekt B: HORA-Kumul (HORA II). Technical Report, IAWG Ottobrunn (2006)

ETC-Based Traffic Telematics

Utilizing Electronic Toll Collection Systems as a Basis for Traffic Data Generation

Thomas Stranner, Peter Ummenhofer, and Alexander Abl

Abstract. The basis for many ITS (Intelligent Transport Systems) applications is high quality data which, up until now, has been captured using dedicated traffic sensor networks. In more and more countries, Electronic Toll Collection (ETC) systems are being implemented to finance the traffic infrastructure and regulate traffic demand. Such systems offer an ideal source of data and allow the use of ETC infrastructure like overhead gantries for further traffic telematics solutions. Therefore ETC systems will increasingly become the "*backbone*" for the telematics strategies of road operators in the future.

1 Introduction

The use of an electronic toll collection system, beyond its primary purpose (automated toll collection), allows the road authorities to improve traffic management and planning, to increase road safety and public security, to improve service offers for the road user and to generate additional revenue through value added services.

Thomas Stranner
Solution Architect Traffic Telematics Solutions,
Kapsch TrafficCom\ AG, Lakeside B03, 9020 Klagenfurt, Austria
Tel.: +43 (0)50811 7833
e-mail: thomas.stranner@kapsch.net

Peter Ummenhofer
Head of Competence Center Traffic Telematics Solutions,
Kapsch TrafficCom AG, Lakeside B03, 9020 Klagenfurt, Austria
Tel.: +43 (0)50811 7810
e-mail: peter.ummenhofer@kapsch.net

Alexander Abl
Solution Architect Traffic Telematics Solutions,
Kapsch TrafficCom AG, Lakeside B03, 9020 Klagenfurt, Austria
Tel.: +43 (0)50811 7834
e-mail: alexander.abl@kapsch.net

J. Düh et al. (Eds.): Data and Mobility, AISC 81, pp. 71–81, 2010.
springerlink.com © Springer-Verlag Berlin Heidelberg 2010

Traffic management, for example, requires up-to-date, network-wide traffic data, which is currently collected via a dedicated infrastructure (e.g. inductive loop sensors). The erection, operation and maintenance of this infrastructure is associated with high costs and, as a result, up-to-date traffic data is frequently not sufficiently available. The use of the toll system for the network-wide capture of traffic data for traffic management as well as for traffic planning and the associated reduction in the number of dedicated sensor points offers a major savings potential.

This paper offers some insight to the Kapsch Telematics Platform, which is a modular software system for implementing secondary ITS applications on the basis of Kapsch toll systems. The platform is in use in the Czech Republic enabling the Czech road authority to utilize the toll system to gather traffic data and provide this as input for traffic management as well as traffic planning.

This paper provides an overview about this new approach for capturing traffic data and describes the various traffic data applications of the Kapsch Telematics Platform, namely a traffic monitoring application allowing the calculation of travel times and level-of-service, a traffic statistics application offering a broad range of 'out of the box' traffic statistics, as well as a traffic flow analysis application allowing the analysis of vehicle routes equipped with an onboard tolling unit.

2 The Need for Traffic Telematics

The current traffic forecast at a European level details expectation for an average increase in the traffic performance in Europe of about 2% per year for the next 20 years. A closer look shows that the increase along major sections of the transport network (e.g. the high-level network) will see an even higher average growth rate of up to 4%. Such sections of the transport network have in many cases already reached the limit of their current capacity. The further expansion of these sections is difficult to finance, problematic for environmental reasons, and very often simply rejected by the population. Thus, new approaches and concepts for the optimization of the transport system will have to be found. In this context, the need for traffic telematics becomes obvious.

The use of information and communication technologies to help manage traffic and provide real-time traffic information to the driver is one of the most important measures that can be taken to counter the negative effects. In addition, data from telematics applications can be used as the basis for traffic planning in order to expand the traffic infrastructure in line with demand.

As well as this, traffic enforcement makes a significant contribution to traffic safety and public security. Speed monitoring reduces the number of accidents as well as pollutants and noise emissions. Automated number plate recognition and vehicle tracking – e.g. for the monitoring of hazardous goods transportation – are further examples of how telematics solutions can be used to cope with the increase in road traffic.

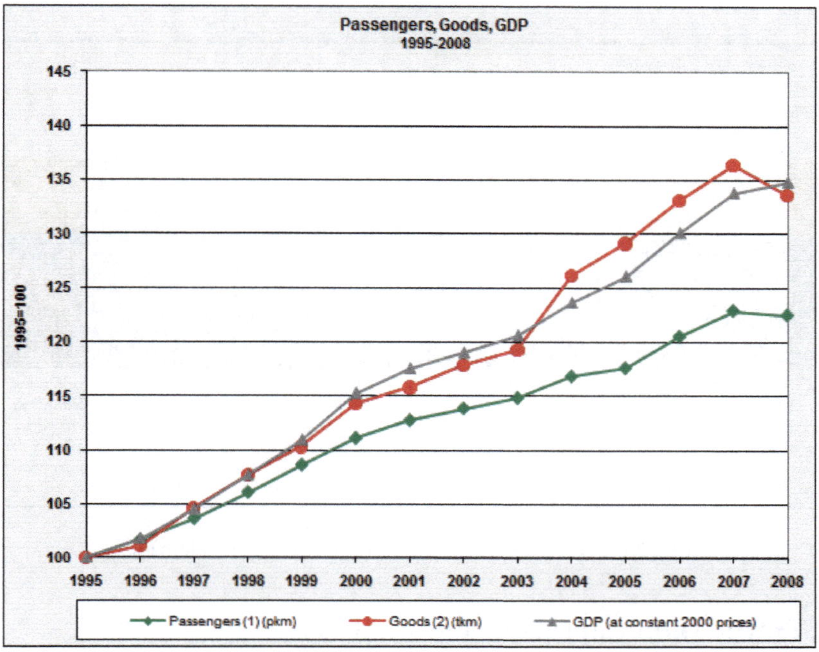

Fig. 1 Transport growth in EU 27, evolution 1995-2008 [1]

3 Electronic Toll Collection Systems as a "Telematics Backbone"

Electronic toll collection systems offer an ideal infrastructure that can be used as the basis for many other telematics solutions in addition to tolling. ETC systems will therefore increasingly become the "*backbone*" for the telematics strategies of road operators in the future· [2]

For example, the data generated by DSRC systems can be used to produce traffic data. Furthermore, the roadside equipment of DSRC systems (toll gantries) can be used for additional purposes such as the installation of traffic data sensors or variable message signs, so that costs for roadside infrastructure can be shared with other applications.

Use of the toll systems as the basis for secondary telematics services saves costs, offers the opportunity for additional revenue sources and contributes to the acceptance of the road user charges among drivers through improved traffic flow, increased safety as well as additional services. Fields of application for ETC-based telematics solutions are traffic management, information and planning, safety and security, mobility services for the road user, as well as various industries.

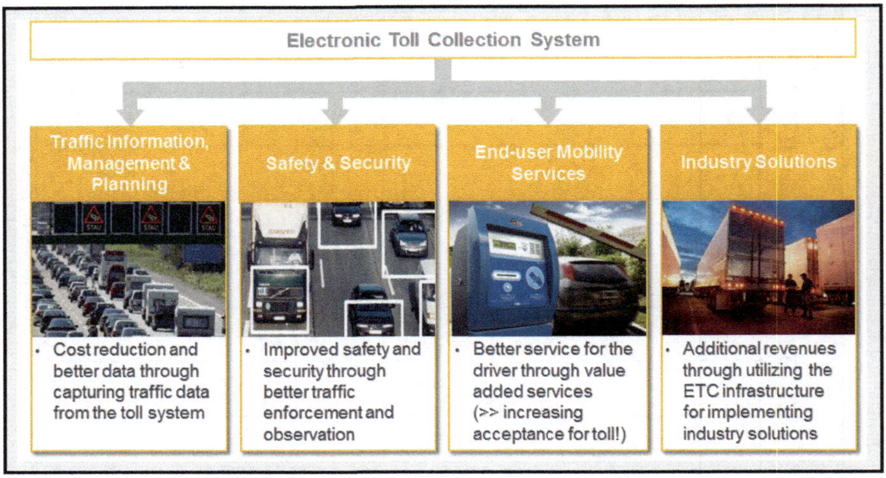

Fig. 2 Field of applications for ETC-based traffic telematics

4 Traffic Data Capturing and Analysis

Traffic management requires up-to-date, network-wide traffic data, which is currently collected using dedicated infrastructure (e.g. inductive loop sensors). According to the study "Telematics Applications in Road Transportation – Status and Perspectives" [3], one of the main problems in traffic control is the seamless installation of detection technology throughout the road network. According to the study, the high resolution collection of traffic data is the greatest guarantee of useful results in traffic-related data processing.

The erection, operation and maintenance of this infrastructure is associated with high costs and, as a result, up-to-date traffic data is frequently not sufficiently available. The use of the toll system for the network-wide recording of traffic data for traffic management as well as for traffic planning and the associated reduction in the number of dedicated sensor points offers a major savings potential. The infrastructure at the toll and enforcement stations (overhead bridges, power supply and data connection) can be used for the installation of traffic data sensors.

In addition, it is possible to generate additional traffic data directly based on the toll system by appropriately processing and sharing toll data so that additional information is available for traffic management and traffic planning.

With the introduction of a toll for both trucks and passenger vehicles and thereby the obligatory equipping of all vehicles with an onboard unit (OBU), dedicated traffic sensors can even be largely eliminated as the all important traffic data can be derived directly from the toll data, therefore further increasing the potential for savings.

4.1 Sample Application 1: Traffic Monitoring

The traffic monitor is used to calculate the actual traffic situation on the road network. This application calculates travel times and the average speed on the basis of DSRC and video toll transactions generated by vehicles passing two toll stations. On the basis of the determined travel times in relation to the base travel time, the application derives level-of-service information. The level-of-service (LoS) is a qualitative measurement describing the operational conditions of a segment or traffic stream. The LoS information can subsequently be made available to external systems like traffic management systems or can be directly presented to road users via a web portal or a wireless web application.

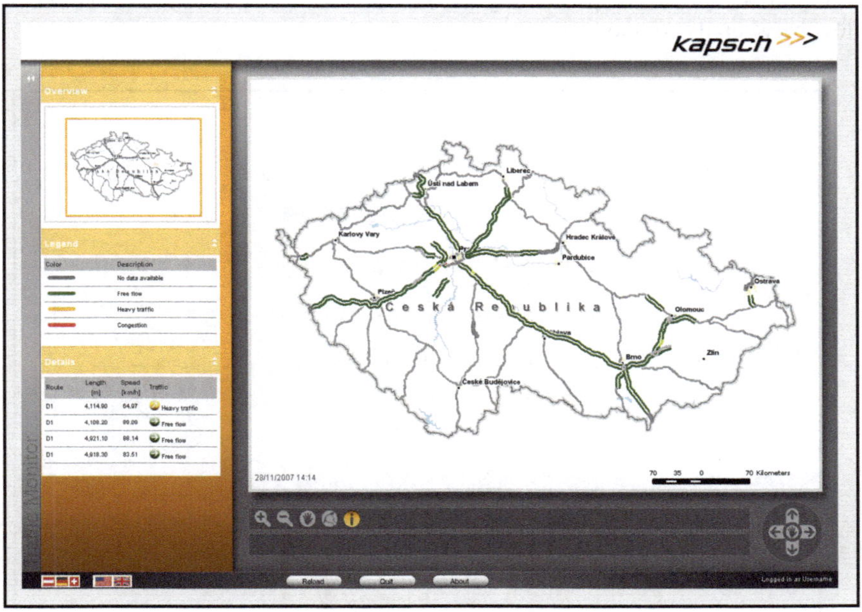

Fig. 3 Example screenshot of Traffic Monitor

Depending on the actual travel times compared to a pre-defined base travel time, the LoS is derived for the analyzed road sections. The base speed can be set manually. In addition, it is possible to automatically calculate the base speed on the basis of historical speed data (typically these values are captured at night time when there is little traffic and therefore an undisturbed traffic situation). A base travel time may also be predicted by traffic planners considering, in addition to traffic parameters, geographical conditions like the road layout or base slope. In such a case, the LoS can be used to evaluate whether the predicted quality could be reached under real conditions.

The level of service is typically visualized as color codes representing the monitored status of the traffic flow: green represents free flow, yellow represents queuing and red represents congestion. The calculation is based on the normal speed and averaged travel times.

Processing Steps

The application is designed as strict component model. Each component is responsible for fulfilling one single step of the process, beginning with data retrieval. Each processing step can be configured independently to ensure that a component can be tuned without interference from the other steps. The configuration data is stored together with the application metadata in the Kapsch Telematics Platform database.

Figure 4 explains the processing steps of the traffic monitor system and the data flow between its components:

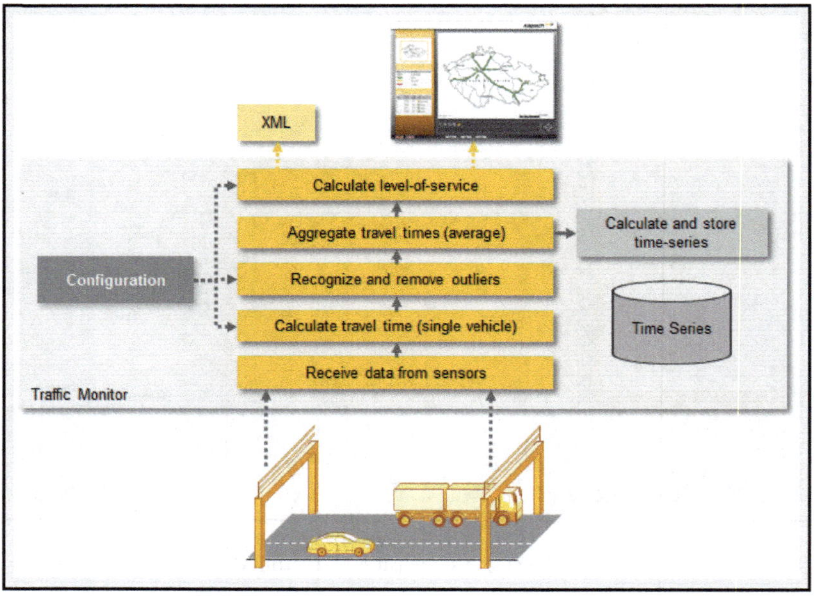

Fig. 4 Process steps of the traffic monitor application

The traffic data gained from the ETC system as the traffic sensor backbone is continuously collected in a repository within the Kapsch Telematics Platform (KTP). At a pre-defined interval, the data is accessed by the travel time calculation module. This module calculates the travel time between two subsequent toll stations representing one segment for each detected vehicle and vehicle class.

A following filter mechanism detects and removes outliers and unreliable samples. The filter method *median* and the *median absolute* deviation have shown optimized results compared to *mean* or *standard* deviation. [4]

$$\tilde{x} = \begin{cases} x_{\frac{n+1}{2}} & ,if\ n\ odd \\ \frac{1}{2}\left(x_{\frac{n}{2}} + x_{\frac{n}{2}+1}\right) & ,if\ n\ even \end{cases}$$

The *median* $\tilde{x}$ of an ordered sample (x_1, x_2, …, x_n) of n measured values is selected as
The *median absolute* deviation is defined as

$$MAD = median\,|x_i - \tilde{x}|$$

Figure 5 shows a time series plot for a road section in Czech Republic. The plot visualizes the increase in travel time for this road section due to a traffic accident at about 11:00 am and the decrease to base travel time after the road was cleared again.

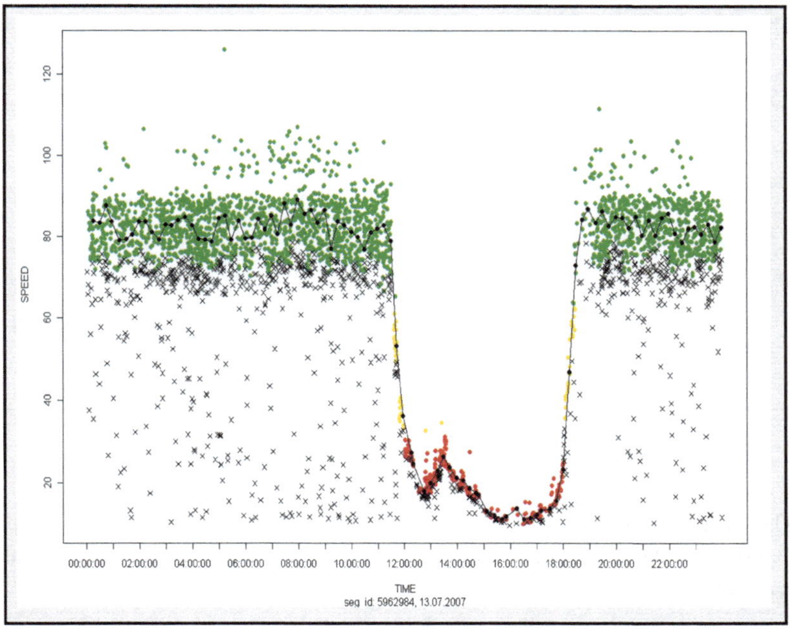

Fig. 5 Time series plot of traffic monitor travel time calculation

The black 'x' represents the filtered outliers and the connecting line between the black points shows the resulting LoS when the calculation interval is 15 minutes. In addition to the data filter, exponential smoothing is used to ensure that the system is not fluctuating. This method – compared to the simple arithmetic mean value – is useful for recognizing trends in succeeding samples. The result of the exponential smoothing is then used to determine the level of service.

4.2 Sample Application 2: Traffic Statistics

This application provides statistics relating to traffic volume and its development over time. The statistics form the basis for planning the transportation infrastructure and the design of traffic-controlling measures. Typical statistics are the amount of traffic per day, vehicle distributions with differentiation between passenger vehicles and trucks during peak hours and monthly peaks based on the day of the week. From one perspective, the statistics are based on the DSRC and video toll data stored in the Data Warehouse of the tolling system as well as in long-term data captured by the vehicle classification devices of the enforcement stations.

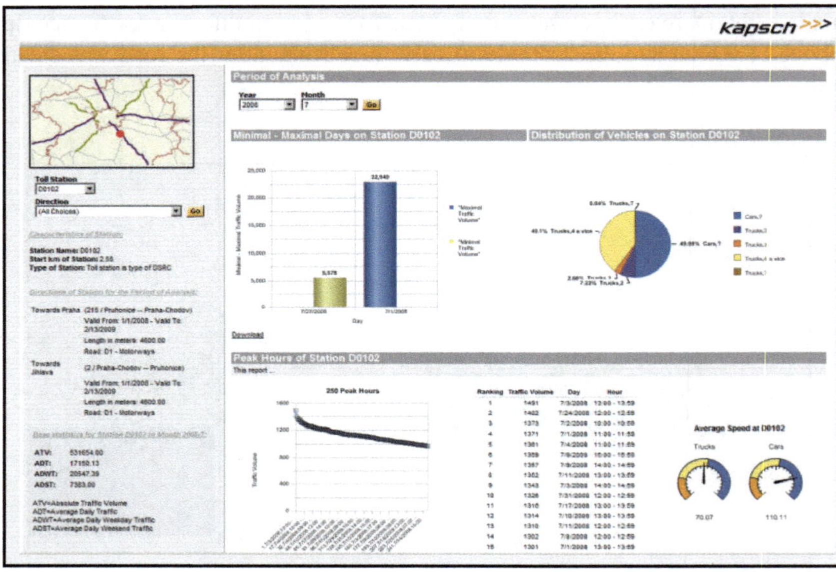

Fig. 6 Report of time series for absolute traffic volume

The Traffic Statistics application is able to process data from multiple sources. Depending on the data source, the parameters which can be chosen for analyzing the data differ in:

- Statistics based on DSRC toll data allow the taking into account of parameters stored in the OBU (e.g. number of axles, emission class etc.).
- Statistics based on video toll data allow the taking into account of parameters provided by the video system (e.g. vehicle length).
- Statistics based on traffic data captured by the laser scanner of enforcement stations allow the taking into account of parameters for long-term traffic data (e.g. vehicle class).

4.3 *Sample Application 3: Traffic Flow Analysis*

A traffic flow analysis allows the analysis of traffic flow within the tolled road network. The basis is route information for single vehicles. Routes of vehicles are determined on a basis of sequences of DSRC or video toll transactions (station passages). There is broad spectrum of circumstances in which the application can be used. Examples are the analysis of the traffic passing a certain station (e.g. "Where do vehicles go to after passing a station?") or the flow between stations (e.g. "How many vehicles have driven from station A to station B and which intermediate stations have they passed?"). Based on the traffic flow analysis, authorities can carry out road network planning and optimization or on evaluation of further toll station locations for example.

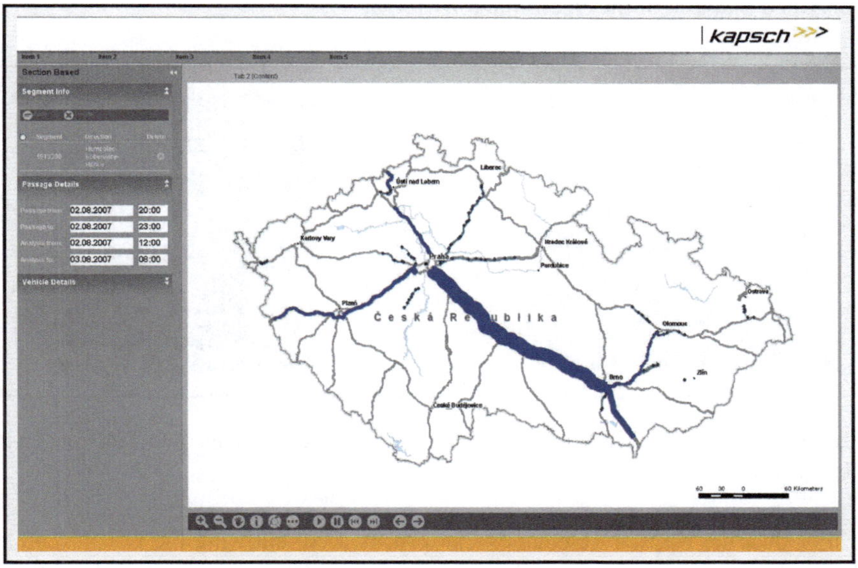

Fig. 7 Traffic Flow Analysis – segment based

Various types of traffic flow analysis are possible by choosing different input parameters. The three main methods for analysis are the segment-based analysis (evaluating a single toll segment), the route-based analysis (routes defined by the user) and the detour analysis (providing the amount of vehicles which have not driven on the tolled network).

5 Case Study: Czech Truck Tolling System

On 1st January 2007, the Czech Republic's nationwide electronic toll collection system "MYTO CZ" entered into commercial operation. One year later, the system had expanded to cover 1180 km of selected highways and motorways, while around 375,000 On Board Units (OBUs) had been registered.

The Czech Government also needed to safeguard and maximize the return on its investment by ensuring the system could be adapted to support future developments without compromising its existing functionality. This might include, for example, adding vehicle classes, and supporting wider traffic management, road safety and other telematic functions. The system has already been augmented to extend its role as a traffic management and traffic planning tool, demonstrating the soundness of an approach based on investing in a solid, revenue-generating system.

Based on the DSRC toll data from the ETC system, traffic data is being obtained by RSD (Czech Road Authority) cost efficiently. For traffic management and traffic planning purposes, accurate real-time as well as historical traffic data is generated on basis of the electronic toll system.

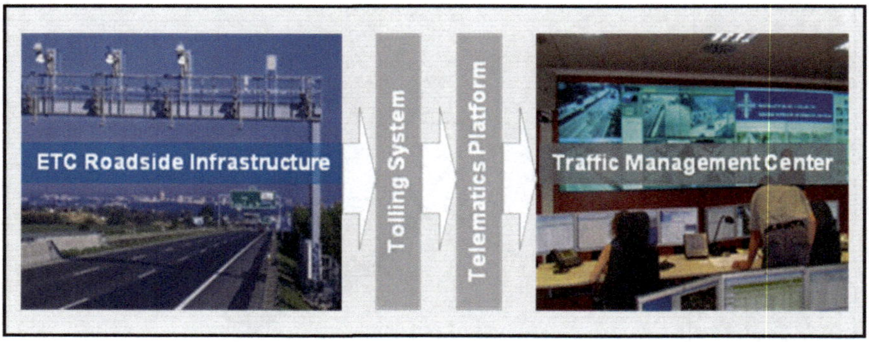

Fig. 8 Data flow of generating real time traffic data based on toll data for Traffic Management

For this, the Kapsch Telematics Platform is used, allowing the Czech road authority to capture travel times, level-of-service, traffic statistics as well as traffic flow analysis data for the entire tolled road network in a cost efficient manner and feed this data via open XML interfaces to the Czech National Traffic Management Center. There, the data is used for controlling traffic and planning traffic infrastructure in line with demand.

6 Conclusion

The success story of the Telematics Platform in the Czech Republic substantiates that an established ETC system can be used as a cost efficient data source for many ITS solutions and it can provide additional data for traffic management and planning purposes. While truck tolling is currently often implemented as a first step, passenger car tolling is increasingly being discussed by governments as an additional source of financing costs related to traffic.

In future scenarios where passenger cars could be charged with electronic tolling systems, traffic data could be generated for all vehicles, allowing a reduction in the number of dedicated traffic sensors used for traffic management and traffic analysis purposes. In addition, road-user oriented traffic telematics solutions like information services, automated access control at parking facilities or cashless payment at gas stations will probably be introduced, thus enabling an increase in comfort and safety for the driver and – in the event that these mobility services are offered by the road authorities – to help increase acceptance for road user charges.

References

1. European Commission: EU energy and transport in figures. Publications Office of the European Union, Luxembourg (2010) doi: 10.2768/19814, ISBN 978-92-79-13815-7
2. Ummenhofer, P.: ETC-Based Traffic Telematics – Utilizing Electronic Toll Collection Systems as basis for value adding Telematics Applications. In: ITS Europe 2008, Geneva (2008)
3. Busch, F., Keller, H., Hoyer, R., Reupke, H., Riegelhuth, G., Zackor, H.: Telematikanwendungen im Straßenverkehr – Stand und Perspektiven Part 1and 2. Journal Straßenverkehrstechnik version 6 and 7, Germany (2004)
4. Hoaglin, D.C., Mosteller, F., Tukey, J.W.: Understanding Robust and Exploratory Data Analysis, pp. 404–414. John Wiley & Sons, Chichester (1983) ISBN 0-471-09777-2

Extraction of Visual and Acoustic Features of the Driver for Monitoring Driver Ergonomics Applied to Extended Driver Assistance Systems

H.D. Vankayalapati, K.R. Anne, and K. Kyamakya

Abstract. The National Highway Traffic Safety Administration (NHTSA) estimates that in the USA alone approximately 100,000 crashes each year are caused primarily by driver drowsiness or fatigue. The major cause for inattentiveness has been found to be a deficit in what we call in this paper an extended view of ergonomics, i.e. the "extended ergonomics status" of the driving process. This deficit is multidimensional as it includes aspects such as drowsiness (sleepiness), fatigue (lack of energy) and emotions/stress (for example sadness, anger, joy, pleasure, despair and irritation). Different approaches have been proposed for monitoring driver states, especially drowsiness and fatigue, using visual features of the driver such as head movement patterns eyelid movements, facial expressions or all of these together. The effectiveness of the approach depends on the quality of the extracted features, efficiency and the responsiveness of the classification algorithm. In this work, we propose the usage of acoustic information along with visual features to increase the robustness of the emotion/stress measurement system. In terms of the acoustic signals, this work will enlist the appropriate features for the driving situation and correlate them to parameters/dimensions of the "extended ergonomics status" vector. Prosodic features as well as the phonetic features of the acoustic signal are taken into account

H.D. Vankayalapati
Institute of Smart System Technologies, Transportation Informatics Research Group, University of Klagenfurt, Klagenfurt, Austria

K.R. Anne
Department of Information Technology, TIFAC-CORE in Telematics, VR Siddhartha Engineering College, Vijayawada, India.

K. Kyamakya
Institute of Smart System Technologies, Transportation Informatics Research Group, University of Klagenfurt, Klagenfurt, Austria

J. Düh et al. (Eds.): Data and Mobility, AISC 81, pp. 83–94, 2010.
springerlink.com © Springer-Verlag Berlin Heidelberg 2010

for the emotion recognition here. In this paper, a linear discriminant analysis based on a classification method using the Hausdorff distance measure is proposed for classifying the different emotional states. Experimental evaluation based on the Berlin voice database shows that the proposed method results in 85% recognition accuracy in speaker-independent emotion recognition experiments.

Keywords: visual features, acoustic features, ergonomics, driver assistance system.

1 Introduction

Today, it is well known that driver drowsiness, possibly compounded by high workloads and the stress of the ever-increasing complexity of surrounding car and traffic environments, is a major cause of both driver stress and discomfort and of severe accidents. Statistics show that over the past couple of decades, the majority of the accidents on European roads have been due not only to poor vehicle technical conditions but also primarily inattentiveness of the driver [1]. The major cause of inattentiveness has been found to be a deficit in what we call in this project an extended view of ergonomics, i.e. the "extended ergonomics status" of the driving process. This deficit is multidimensional as it includes aspects such as drowsiness (sleepiness), fatigue (lack of energy) and emotions/stress (for example sadness, anger, joy, fear, boredom, neutral state of mind and disgust). In order to improve the "extended ergonomics status" during driving, the European Union (EU) has introduced regulations which include driving time and rest periods for drivers [2]. Even though these regulations have partially helped to improve driver attentiveness, they largely ignore the real-time dynamic "extended ergonomics status". Therefore, there is a need for an emotion recognition system, ideally a non-intrusive one, which can be used in extensive studies, by appropriate professionals, for studying the exact impact of the real-time spatio-temporal dynamics of the "extended ergonomics status" on road safety in general.

Driver behavior changes based on ergonomics which is influenced by diverse factors such as traffic control strategies, road geometry, vehicle characteristics, changing traffic scenarios, weather, etc. Different approaches have been proposed for monitoring driver states, especially drowsiness and fatigue. A first class of approaches looks at vehicle performance/behavior in order to indirectly infer driver status [3-8]. However, these indirect approaches heavily depend upon vehicle and road conditions (e.g. quality of lane markings, alternate lane markings during road repairs) as well as on environmental conditions (e.g. shadow, rain and night vision). These drawbacks have drawn the researcher's interest the direct monitoring of driver behavior. Thus, a second class of approaches directly measures driver physiological characteristics but in an intrusive way by involving measurement systems such as the Electroencephalogram (EEG) which monitors brain activities, the Electrocardiogram (ECG) which measures heart rate variation, the Electrooculogram (EOG) which monitors eye movement, skin potential level measurement techniques, etc. [9-10]. The methods in the second class of approaches require the driver's cooperation as the electrodes are attached directly to the driver's body. Due to the expected limited user acceptance of these intrusive methods in normal vehicles, they are more realistic for a daily use in health care or similar special vehicles only. A further problem is that the intrusive apparatus involved in these methods may itself contribute to driver distraction and fatigue. A third class of approaches is better due to being non-intrusive. Non-intrusive approaches generally involve machine vision as an alternative to a direct measurement of physiological characteristics and they do not need any cooperation from the driver; they monitor driver behavior and status directly through visual sensors. Video sensors are placed on the dashboard to measure, for example, eyelid movements (open/close interval of eyelid), head movements, mouth movements (yawning) and facial expression measures [1, 3, 10-12].

Feature extraction based on visual and acoustic information from the driver is an important and basic task in determining driver behavior/emotions. Generally, a feature is a set of measurements. Each measurement contains a piece of information, and specifies the property or characteristics of the object present in the given input. In the daily life, humans are able to guess and understand the state of others by observing multiple features such as body language, voice information and the interpreted knowledge of understanding what he/she is saying and how. Observing and extracting multiple features is a less complex task (in some cases it is obvious) for humans whereas for machines it is much more complex. This work mainly focuses on (a) identifying the useful features in the acoustic information (b) extracting the features in real time without missing the frames and (c) correlating the features to parameters/dimensions of the "extended ergonomics status" vector.

2 Non-intrusive Feature Extraction Approaches

Feature extraction is used to reduce the large input of data into smaller data quantities and converts the data into small features sets or feature vectors (n-dimensional vector to store numerical features which represent an object). Feature extraction is defined as the process of extracting the feature from the source data, where the data can be embedded from a high dimensional data set [11]. We calculate different feature sets for different applications. For computer vision applications, edges and corners are calculated as features for images. Features like data and noise ratio, length of sound and relative power are calculated for pattern recognition applications.

2.1 Feature Extraction from the Visual Information

In the 1990s, researchers introduced appearance-based linear subspace techniques, statistics related techniques, to reduce the dimensionality and to extract the useful visual features. The introduction of the linear subspace techniques is a milestone in the visual feature extraction concept. The performance of appearance-based techniques heavily depends upon the quality of the extracted features from the image [11]. The appearance-based linear subspace techniques extract the global features, as these techniques use the statistical properties like the mean and variance of the image [13]. The major difficulty in applying these techniques over large databases is that the computational load and memory requirements for calculating features increase dramatically for large databases [13]. In order to increase the performance of the feature extraction techniques, nonlinear feature extraction techniques are introduced. In order to improve the performance of the emotion recognition systems, we have to extract both linear and nonlinear features. There are many nonlinear feature extraction techniques, such as radon transform and wavelet transform. The radon transform based nonlinear feature extraction gives the direction of the local features. This process extracts the spatial frequency components in the direction for which radon projection is computed [14]. When features are extracted using radon transform, the variations in this facial frequency are also boosted [14]. The wavelet transform gives the spatial and frequency components present in an image. The performance of these feature extraction approaches is systematically evaluated in our previous work using the FERET database for face recognition application as shown in Fig. 1 [14].

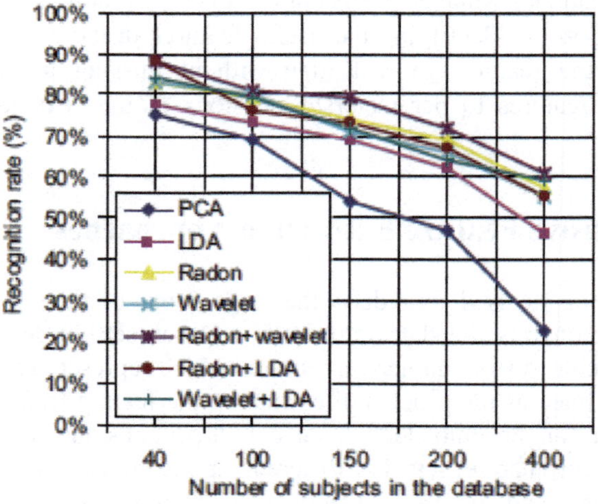

Fig. 1 Performance comparison of different face recognition approaches with profile right images [14]

2.2 *Feature Extraction from the Acoustic Information*

Speech is easy to record, even under extreme environmental conditions (temperature, high humidity and bright light) andmerely requires cheap, durable and maintenance-free sensors and most importantly, it utilizes communication system hardware alerady in existence. Furthermore, speech data is omnipresent in many professional driver settings. Given these obvious advantages, the renewed interest in computational demanding analyses of vocal expressions has been enabled recently given the advances in computer processing speed. The first investigations into emotion recognition from speech were conducted around the mid-1980s. Due to the increased number of speech driven applications in the recent years, the automatic assessment of emotions from the driver's speech signal has become a subject of research [15-16]. Such assessment provides information about the driver satisfaction with the car's infotainment system and in particular, improves the efficiency and friendliness of human-machine interfaces.

For the machine-based state estimation, we will not focus on the voice content but rather on voice-signal features that are relevant for an emotional state inference. In this regard, to make the system more robust in predicting the driver state, we will analyze acoustic information such as pitch, intensity, speaking rate, voice quality, etc. in order to extract the appropriate features [16]. Acoustic features extraction is challenging in many aspects as it depends heavily on the age and gender of the person. The acoustic features vary significantly for different age groups and different genders. Angry males show higher levels of energy than angry females. It is found that males express anger with a slow speech rate as opposed to females who employ a fast speech rate under similar circumstances [16]. Acoustic information will however be very valuable, where available, for better assessing and understanding the effect of driving process ergonomics on the driver's state and mood.

The real-time extraction of acoustic characteristics from the voice signal conveys emotion and attitude in a systematic manner and is different in males and females. Acoustic features are used to recognize the frustration of the driver (i.e. recognize vocal signals through shouting, crying, etc which are not uncommon during the driving process). The performance also depends upon the given (input) acoustic information. In this work, input data is taken from audio sensors like microphones and the feature extraction algorithms are executed to extract the features in real time. Features are extracted from the real time data by performing time and frequency domain algorithms [11]. These algorithms extract temporal, spectral features and cepstral coefficients. These features are extracted based on the amplitude and spectrum analyzer of the audio data. The basic approach to the extraction of acoustic features is frame blocking such that a stream of given input audio signal is converted to a set of frames. The time duration of each frame is about 10~30ms. If the frame size is shorter than 10ms, we may miss some important information and sometimes we cannot extract valid acoustic features. If the frame size is longer than 30ms, redundancy may occur and we cannot capture the time-varying characteristics of the audio signals.

The important voice features to consider for emotion classification are: Fundamental frequency (F0) or Pitch, Intensity (Energy), Speaking rate, Voice quality and many other features that may be extracted/calculated from the voice information such as formants, vocal tract cross-section areas, MFCC (Mel Frequency Cepstral Coefficient), Linear frequency cepstrum coefficients (LFCCs), Linear Predictive Coding (LPC) and the Teager energy operator-based features [16]. Pitch is the fundamental frequency of audio signals, which is equal to the reciprocal of the fundamental period. It can also be defined as the highness or lowness of a sound.

$$F0 = 1 / T0$$

Generally for pitch estimation, wavelet transforms are used. The shape of the vocal tract is modified depending up on the emotion [17]. The MFCC is the "spectrum of the spectrum" used to find the number of voices in speech. It has proven beneficial in speech emotion detection and speech detection tasks. The teager energy operator is used to find the number of harmonics due to a nonlinear air flow in the vocal tract [18-19]. The LPC provides an accurate and economical representation of the envelope of the short-time power spectrum. This is one of the most powerful speech coding analysis techniques which provides very accurate estimates of speech parameters and is known to be relatively efficient for computation at the same time. The LFCC is similar to MFCC but without the perceptually oriented transformation into the Mel frequency scale; it emphasizes changes or periodicity in the spectrum, while being relatively robust against noise. These features are measured based on the mean, range, variance and transmission duration between utterances [15-16]. To calculate these voice features, different techniques are used. The variation in the feature set for joy and sadness are shown in Fig. 2 and Fig. 3 respectively.

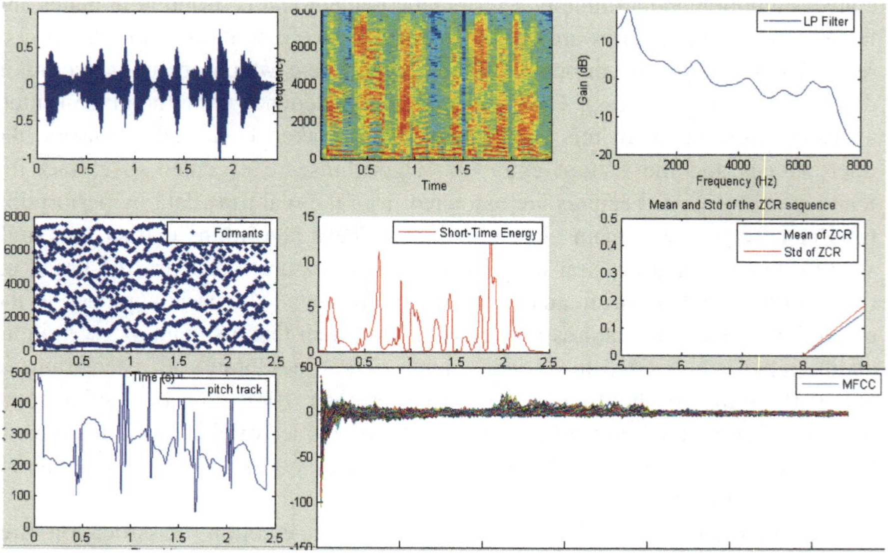

Fig. 2 Extracted features from the joy emotional audio file

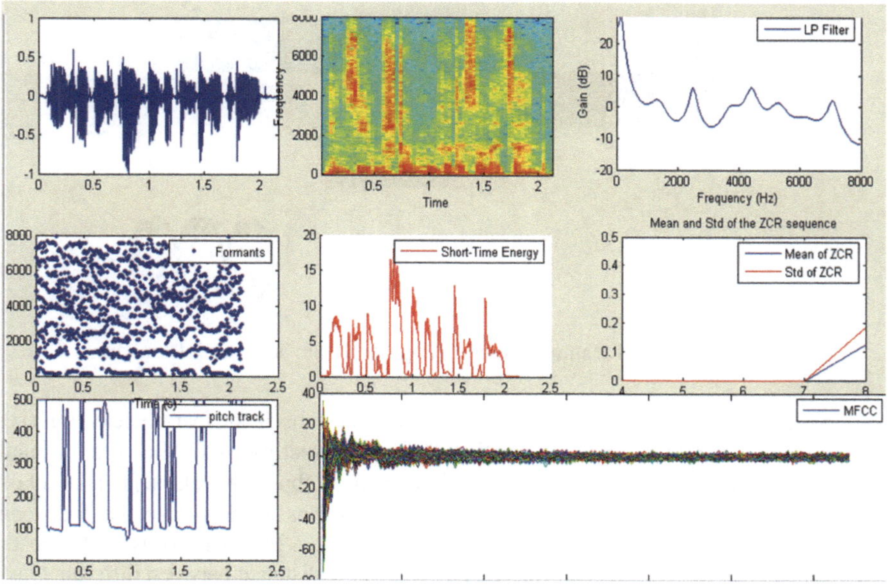

Fig. 3 Extracted features from the sad emotional audio file

2.2.1 Emotion Classification

Emotion detection has been implemented based on a variety of classifiers including Fisher's linear discriminant analysis (FLDA), maximum likelihood classifier (MLC), neural network (NN), k-nearest neighbor (k-NN), and Gaussian mixture model (GMM) [16]. The main criterion in evaluating the effectiveness of the classification algorithm in this work is the scalability of the classifier. By considering the importance of scalability of the classification algorithm, the algorithms are evaluated using the open source Berlin database. For validation purposes, we create the test and training set databases using Berlin Database [20]. A training set database contains:

- 535 utterance speech recordings
- 10 actors: 5 males, 5 females
- 10 utterances (in German) for each voice
- seven different emotional speeches (anger, joy, neutral state, boredom, fear, sadness and disgust) [20]

The test database contains different emotions of the persons present in the training database. The backend classifier works using a fixed length feature vector and is used for classification as shown in Fig. 4.

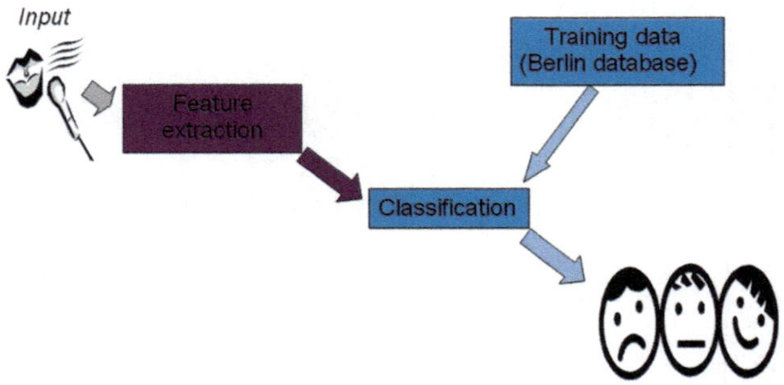

Fig. 4 The overall feature extraction approach from acoustic information

The Linear discriminant analysis (LDA) classifier is used for feature selection (i.e. converts the variable large dimensional feature vectors into the fixed small dimensional feature vector). LDA uses information about the class [11]. A class contains one person with different emotions. LDA tries to maximize the 'between class' variance and minimize the 'within class' variance. In other words, it decreases the distance between same class files and increases the distance between different class files [11]. Because of this, LDA easily recognizes the emotions in large databases. The success rates of the different popular classifiers are compared in Table 1.

Table 1 Major classification techniques and their success rates

Classification techniques	Success rate
Linear discriminant analysis (LDA)	67.22%
k-nearest neighbor (k-NN)	63.33%
neural network (NN)	57.78%
maximum likelihood classifier (MLC)	55%
Gaussian mixture model (GMM)	53.33%

2.2.2 Proposed Classifier

The LDA performs considerably better when compared to the above classifiers for the Berlin database. But the performance of LDA is insufficient for real world applications.

- The scalability of the real world emotion recognition system is limited, as the computational load and memory requirements increase dramatically with the large data sets. So scalability is a major issue here.
- Up to now, in LDA we have been using linear metric (Euclidean distance). Linear metric cannot compare different dimensions of the acoustic feature vectors accurately.

To improve the performance (success rate and process speed), we propose the nonlinear Hausdorff metric-based LDA. A special nonlinear metric Hausdorff distance is able to compute the distance between different sized matrices having a single common dimension, like the acoustic matrices representing our acoustic feature vectors [21-22].

Hausdorff distance measure is often used in content-based retrieval applications [21-22]. Hausdorff distance is intended to be a measure between two point collections A and B in a metric space S (whose distance is d); it can be viewed as a dissimilarity measure between two feature vectors A and B. By considering the Hausdorff distance measure instead of the linear Euclidean distance measure, the success rate of the LDA algorithm is increased by around 20 % as shown in Fig. 5.

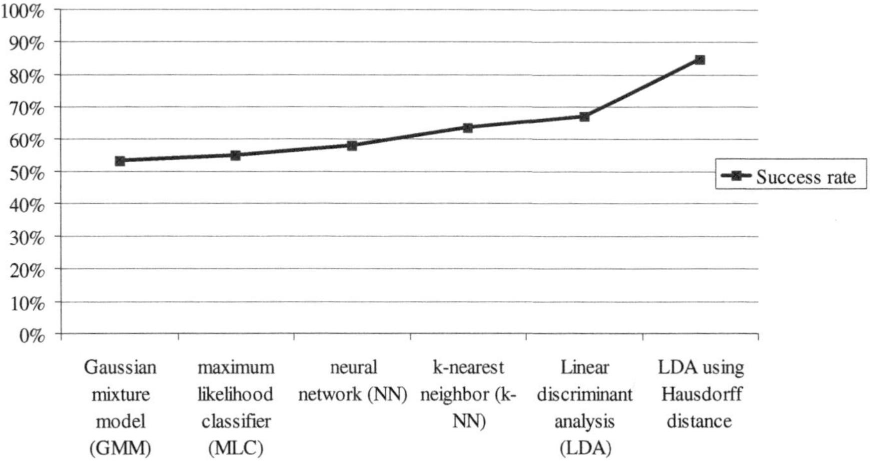

Fig. 5 The graphical representation of the success rate for different classifiers

3 Conclusion

The emotion recognition performance using acoustic information generated by the driver is systematically evaluated by using the Berlin emotional database. Emotion detection has been implemented based on a variety of classifiers including linear discriminant analysis (LDA), maximum likelihood classifier (MLC), neural network (NN), k-nearest neighbor (k-NN), and the Gaussian mixture model (GMM). Among these algorithms, linear discriminant analysis performed much better with 67.22%. In this work, we proposed the application of the Hausdorff distance measure instead of the Euclidian distance measure to extract the nonlinear features. The results of the evaluation have shown that the success rate increased by over 20% with the combination of LDA and Hausdorff distance.

4 Future Work

In this research, we intend to use multi-modal features from the driver to be extracted from three different sensor systems, namely visual, thermal and acoustic as shown in Fig. 6 to classify the extended ergonomic state of the driver.

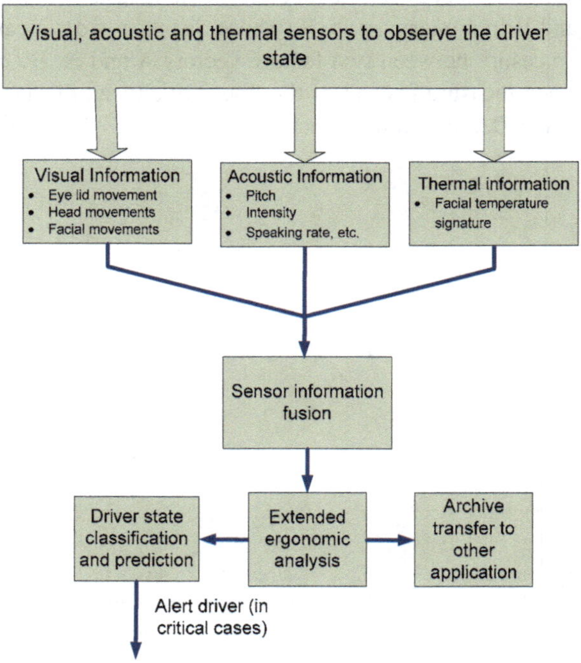

Fig. 6 The global concept description of the proposed "extended ergonomics status" monitoring system

The features extracted from the visual, acoustic and thermal information will be fused together and the result integrated into a state or "extended ergonomics status" prediction model involving concepts such as dynamic Bayesian networks or artificial neural networks, etc. The overall system determines the driver's "extended ergonomics status" vector which includes components such as fatigue, drowsiness and emotions (such as anger, despair, pleasure, sadness, irritation, joy). The "extended ergonomics status" vector is also beneficial to highway Police in analyzing the causes of accidents.

References

1. Boverie, S.: Driver fatigue monitoring technologies and future ideas. Presented at the AWAKE Road Safety Workshop, Balocco, Italy (2004)
2. EU, Regulation (EC) No. 561-2006 of the European parliment and of the council of 15th march 2006 on the hramoniyation of certain social legislation relating to road transport and amending council regulations (EEC) No. 3821-85 and (EC) No. 2135-98. Offical Journal of the European Union, 102 (2006)
3. Fletcher, A.L., Chen, J., Zelinsky, A.: Computer vision for vehicle monitoring and control. Presented at the Australian Conference on Robotics and Automation, Sydney (2001)
4. Apostoloff, A.Z.N.: Robust vision based lane tracking using multiple cues and particle filtering. Presented at the University of Oxford, Oxford, UK (2003)
5. Onken, R., Feraric, J.P.: Adaptation to the driver as part of driver monitoring and warning system. Accident Analysis and Prevention 29, 507–513 (1997)
6. Pilutti, T., Ulsoy, A.C.: On-line identification of driver state for lane-keeping tasks. In: 1995 American Control Conference, Seattle, Washington, pp. 678–681 (1995)
7. Pilutti, T., Ulsoy, A.C.: Identification of driver state for lane-keeping tasks. IEEE Transactions Systems, Man and Cybernetics, Part A, 486–502 (1999)
8. Wijesoma, W.S., Kodagoda, K.R.S., Balasuriya, A.P.: Road boundary detection and tracking using ladar sensing. IEEE Transactions on Robotics and Automation 20, 456–464 (2004)
9. Saroj, P.B.A.C., Lal, K.L., Kirkup, L., Nguyen, H.: Development of an algorithm for an EEG based driver fatigue countermeasure. Journal of Safety Research 34, 321–328 (2003)
10. Roman Bittner, L.P.K.H., Smrcka, P., Schreib, P., Vysoký, P.: Detecting of Fatigue States of a Car Driver, London, Uk (2000)
11. Wahlstrom, O.M.E., Papanikolopoulos, N.: Vision-based methods for driver monitoring. In: Proceedings of the 6th IEEE International Conference on Intelligent Transportation Systems, Shanghai, China, pp. 903–908 (2003)
12. Hyun Jae Baek, J.S.K.H.B.L., Choi, J.M., Kim, K.K., Park, K.S.: Nonintrusive Biological Signal Monitoring in a Car to Evaluate a Driver's Stress and Health State. Telemedicine and e- Health 15, 182–189 (2009)
13. Esra Vural, A.E.M.C., Littlewort, G., Bartlett, M., Movellan, J.: Drowsy driver detection through facial movement analysis. In: Lew, M., Sebe, N., Huang, T.S., Bakker, E.M. (eds.) HCI 2007. LNCS, vol. 4796, pp. 6–18. Springer, Heidelberg (2007)
14. Vankayalapati, H.D., Kyamakya, K.: Nonlinear Feature Extraction Approaches for Scalable Face Recognition Applications. ISAST Transactions on Computers and Intelligent Systems 2 (2009)
15. Bezooijen, R.V.: The Characteristics and Recognizability of Vocal Expression of Emotions, Foris, Drodrecht, The Netherlands (1984)
16. Thurid Vogt, J.W.E.A.e.: Automatic Recognition of Emotions from Speech: a Review of the Literature and Recommendations for Practical Realisation. In: Affect and Emotion in Human- Computer Interaction: From Theory to Applications Berlin (2008)
17. Murray, J.A.I.: Toward the simulation of emotion in synthetic speech: A review of the literature on human vocal emotion. Journal of the Acoustical Society of America 93(2), 1097–1108 (1993)

18. Ververidis, C.K.D.: Emotional speech recognition: Resources, features, and methods. Presented at the Speech Communication (2006)
19. Yongjin Wang, L.G.: An investigation of speech-based human emotion recognition, pp. 15–18 (2004)
20. Paeschke, M.R.A., Sendlmeier, W., Weiss, B.: A Database of German Emotional Speech. In: Proc. Interspeech (2005)
21. Barbu, T.: Discrete speech recognition using a hausdorff based metric. In: Proceedings of the 1st Int. Conference of E-Business and Telecommunication Networks, ICETE, Setubal, Portugal, pp. 363–368 (2004)
22. Barbu, T.: Speech-dependent voice recognition system using a nonlinear metric. International Journal of Applied Mathematics 18, 501–514 (2005)

How Motorcycle Collisions Depend on Weather

Martin Winkelbauer, Christian Brandstätter, Sebastian Riegler, Reinhold Steinacker, and Michael Tiefgraber

Abstract. It is a matter of common sense that motorcycle accidents depend heavily on the actual weather and the weather forecast. This impact may be derived from differences in the intrinsic risk of riding under different weather conditions, differences in the risk-taking behavior of motorbike riders under different weather conditions as well as the impact of weather on exposure. However, such relationships have not been researched thusfar. Traditional analysis of accident statistics will never provide sufficient information on this issue, since weather information in a police report is just about the weather at the very time and location of the collision, but not about the remaining 364 days and 23 hours of a year. Hence, traditional analysis of police reports is more about measuring exposure instead of risk. In addition to this, accident statistics on powered two-wheelers are generally considered to be more informative about the weather in a particular year than about developments in motorcycle safety.

For this study, a particular weather database prepared by the Institute for Meteorology and Geodynamics at the Vienna University was used. This database provides information about precipitation at each corner point of a grid with a cell size of about 16 km x 16 km. This database was linked with the Austrian (police recorded) accident database on a geo-referenced basis, from the year 2002 to 2004, for which the relevant weather information was available. Cross-validation with police-recorded weather data was carried out successfully.

As input for further calculation, a parameter called "rain likeability" was calculated for the whole of Austria. Motorcycle collisions and rain likeability were found to have a close exponential correlation. It was also found that there are different characteristics in terms of accident types and crash severity depending on rain likeability.

Martin Winkelbauer · Christian Brandstätter · Sebastian Riegler
Kuratorium für Verkehrssicherheit (KfV).

Martin Winkelbauer
Kuratorium für Verkehrssicherheit (KfV), Schleiergasse 18, 1100 Vienna
Tel.: +43 (0)5 77 0 77-1214
e-mail: martin.winkelbauer@kfv.at

Reinhold Steinacker · Michael Tiefgraber
Universität Wien, Institut für Meteorologie und Geophysik (UNIVIE)

J. Düh et al. (Eds.): Data and Mobility, AISC 81, pp. 95–106, 2010.
springerlink.com © Springer-Verlag Berlin Heidelberg 2010

These findings were used to review police-reported accident statistics and eliminate the impact of weather. It was found that "predicting" the number of motorcycle accidents in a particular year is possible with an error of less than 3% based on the weather characteristics of this particular year alone and the relationship between weather and accidents. This will, in the future, move the analysis of motorcycle accidents much closer towards displaying a true picture of developments in motorcycle safety. This methodology can be applied to improve the analysis of any other activity that is dependent on weather conditions, where events take place at different sites.

1 Introduction

In the course of the European Commission-funded project, a macroscopic analysis of PTW (powered two-wheelers, i.e. mopeds and motorcycles) collisions[1] has been carried out. 3% (Italy, Greece) to 7% (Austria) were found to take place under rainy conditions according to the police recorded accident databases for the respective countries. Unfortunately, this information is hardly worth anything without knowledge about certain factors e.g. the number of rainy days or, in the optimum case, the number of kilometers driven in rain. In other words, to determine the specific risk of driving in rain, it is necessary to have precise exposure data, which is not available at the moment and is very difficult to collect.

However, for other purposes, exposure data is not needed at all, if exposure is considered to be mileage. Exposure may also mean being exposed to certain weather conditions. This approach was chosen for the work described below. The initial idea for this research is derived from phenomena recorded in the Austrian accident statistics. During the last 20 years, the number of licensed motorcycles has quadrupled. Graduated licensing, probationary licensing, a penalty points system, improved driver training, multiphase driver training and many other measures were implemented with the expectation that they would have an impact on the number of PTW rider collisions. Nevertheless, the number of motorcycle rider fatalities throughout these 20 years varied between 80 and 120 per year, without any clear trend. There may be a great deal of explanations for the lack of a general trend, where the vast safety improvements (e.g. suspension, brakes, ABS, tires, environment, training, enforcement) and changes in rider population cancel out the additional exposure. But there is no explanation for the huge variance, except the impact of the weather.

This also means that it is difficult to evaluate the effects of safety measures, since they are covered by the impact of weather. This research was carried out to develop a methodology to eliminate the random impact of weather in order to gain a clearer picture of the real development of PTW collisions.

---

[1] The term "accident" implies that such events are unavoidable and just happen by chance. "Crash" might be an alternative, but "collision" is the best English term and is therefore used in this paper, except for with habitual language uses technical terms like "accident database". "Collision" also includes single vehicle events, where occupants and/or vehicles collide with elements of the road infrastructure or simply the road surface itself.

This study considers itself to be a pilot. The literature survey carried out at the beginning found that hardly any studies are conducted on the correlation between weather and accidents and those which are, are not controlled for exposure effects. Hence, due to our limited resources, we have not tried to include all the parameters of weather, and a relatively small sample was accepted. The main purpose was to find out whether the methodology can discover any impacts caused by weather and whether this can be put into mathematical terms.

PTW riders, in particular motorcycle riders, are a good subject of investigation. There are two kinds of motorcycle trips, i.e. those, which are completed just for fun and those which have a particular goal (mainly commuting). It is a matter of common sense that recreational rides are more dependent on the weather than riding for a purpose. In other words, if a motorcycle rider wants to use his motorcycle at the weekend just for the fun of it, he or she will quite likely not take the bike out if rain is expected. On the other hand, PTW riders using their bike for the daily trip to work have a particular purpose in mind, e.g. avoiding getting stuck in a traffic jam, avoiding having to look for a parking place for their car or saving travel time. Such riders might be less likely to refrain from riding due to bad weather.

Besides this, it is quite likely that most of the recreational rides happen on weekends whereas commuting happens during the working week. Hence, different impacts of weather on PTW collision can be expected for different parts of the week. If this were achieved, it would give an additional indication that there is a relation between weather and PTW collisions.

2 Previous Studies

Weather has been reported to be a less influential factor in 98% of motorcycle accidents in research conducted in California (Hurt et al. 1981). In a MAIDS report, weather made no contribution to causing accidents in 92.7% of MAIDS cases (854 cases) and was the precipitating event in 7 cases (0.8% of all cases); weather conditions at the time of the accident were most frequently dry (90%). Rain at the time of the accident was noted in 8% of all cases, whereas dry conditions free of contamination were recorded in 85% of all accidents. More than 80% of crashes entailing the death of a motorcyclist between 1999 and 2003, on Australian roads, were reported under fine weather conditions (Johnston et al. 2008). Riding in fine weather also appears to result in more severe injuries regardless of what control measure was employed (Pai and Saleh 2007).

However, the vast majority of studies dealing with weather and road collisions do not particularly address PTW riding. It is either general information or else focuses on car crashes.

2.1 Meteorosensitivity

Several studies have been carried out to prove that meteorosensitivity is an impacting factor on road traffic risk. Runge (1991) investigated 18,000 traffic accidents in the City of Hamburg, Germany. He found that accidents were 30% more frequent on days which were critical for meteorosensitive people. Beine (1976) identified effects on a theoretical basis by exposing people to electrostatic fields of different intensity. Anderle & Urban (1993) investigated the Austrian accident records for influences caused by the foehn winds. They could not find any direct impact on accident numbers. But they found that the variation in accident numbers is significantly higher on days where there are foehn winds. They concluded that the positive impact of foehn winds on driving conditions (no fog, no rain, no black ice, etc.) cancels out the negative impact on human physiology.

2.2 Road Surfaces

The type and quality of the road surface are certainly a major issue in terms of motorcycle safety. Leden and Salusjärvi (1989) identified more accidents on old than on new road surfaces in a typical before/after study. Within the same research program in the northern European countries, it was found that old roads are somewhat safer than new ones under dry conditions, while it is the other way round during rain (Hemdorff, Leden, Sakshaug, Salusjärvi and Schandersson, 1989).

2.3 Weather Parameters

A study carried out on a road section in Iceland (Sigthorsson, H., Finnsson, S.A., 1997) found increases (40 to 70%) of accident numbers under bad visibility conditions. Snowfall has an even greater effect. Edwards (1998) found road crashes have more severe consequences under adverse weather conditions (rain, fog and high winds). Keay and Simmonds (2005) found that rainfall reduces the traffic volume and increases road accident numbers at the same time. This study used weather information observed every 3 hours. Accident counts increase proportionately to the intensity of rain.

A study by Junk et al. (2005) focused on local correlations between weather and accident counts with particular regard to accidents on wintery road surfaces. This study linked a weather database with an accident database and found some correlations. Similar to Keay and Simmonds (2006), the study found that a short period of rain after a long dry spell causes the highest level of additional risk, which increases with the length of the dry spell. Keay and Simmonds (2006) also found that heavy rain is more hazardous than low rainfall.

However, no study could be found that focuses specifically on PTW collisions. Neither could any study be found which tries to turn weather into a controlled variable in terms of accident statistics.

3 Methodology

3.1 Hypotheses

It is assumed that there are three functions linking the parameters of weather with collision counts:

- the difference in the intrinsic risk of riding under different weather conditions
- differences in the risk-taking behavior of riders under different weather conditions, maybe as a compensation for higher subjective risk on wet roads
- the impact of weather on exposure

This research does not look at the single functions because, whilst this might be interesting to know, it is not necessary. To fulfill the task of mathematically linking weather parameters to accident counts, it is only necessary to have a general function, which includes these three.

3.2 Research Questions

The particular research questions for this work were defined as a set of three hypotheses:

1. PTW collisions correlate with weather conditions. This correlation can be described in mathematical terms.
2. The accident record for a year can be normalized using a correlation between collisions and weather conditions.
3. The correlations between weather and collision numbers differ for weekends and workdays.

Hypothesis 3 refers to the idea that PTW riders do not like driving in rain very much. Hence, recreational rides at the weekend might be refrained from on rainy days. For commuting, other priorities might be relevant. Hence, elasticity of PTW collisions might be lower on workdays than on weekends.

3.3 Weather Database

Weather information in Austria is collected by about 150 stations. They record the following parameters, among others: precipitation, temperature, humidity, wind and air pressure. Although it would be interesting to include all these parameters, it was assumed that precipitation has the strongest impact on PTW riding. The other parameters were therefore not considered in the first stage. Other methods of measuring precipitation would have been possible, such as radar, satellite observation and automatic weather station, all with different spatial and time resolutions. However, data based on human observation every three hours was considered most suitable for the purpose of this study.

Weather stations are not evenly distributed over the country. However, for every potential location of a collision, weather information is needed. The Institute of Meteorology and Geophysics as part of the "MESOCLIM" project has developed the VERA (Vienna Enhanced Resolution Analysis) database, which is based upon ERA-40 raw data. VERA provides weather information for each crossing point on a grid. The side length of the grid is between 10 and 20 km, depending on the parameter (Steinacker 2000, 2006). For this research, a grid of 16 x 16 km for precipitation was used. Intensity of rain is provided as a variable of 6 categories from 0 (no rain) to 5 (heavy rain), see example in Fig. 1.

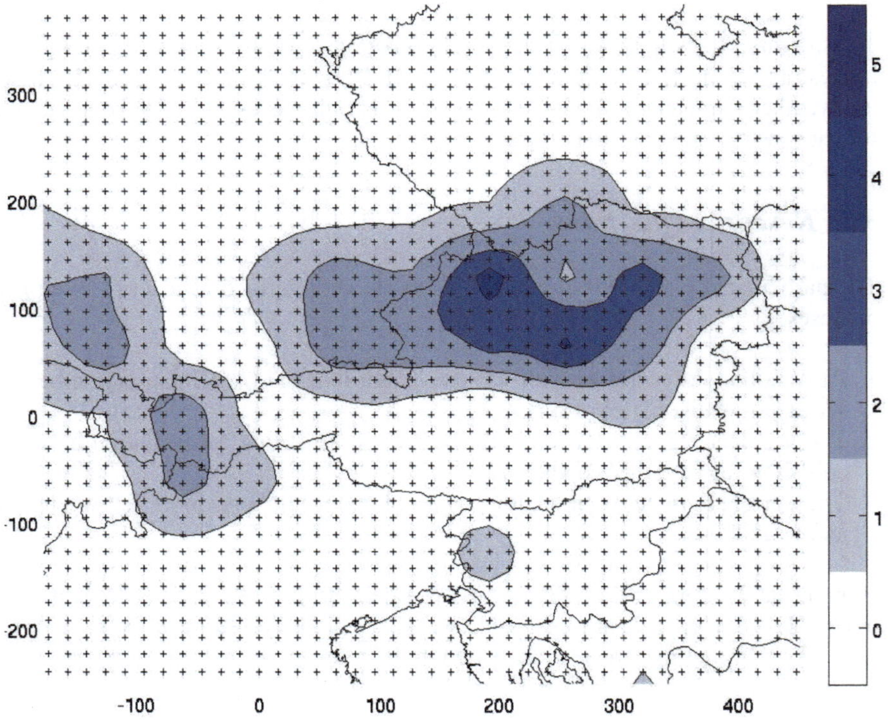

Fig. 1 Example of weather data provided by UNIVIE

Preparing this kind of weather database requires considerable effort - in particular in quality control and validation. Due to particular difficulties with precipitation, this work has, so far only been carried out for other weather parameters. In the end, data for three other years was prepared for further analysis: 2002 to 2004, where 2002 and 2003 data was available initially and 2004 was added later.

3.4 Accident Database

Austrian road accidents are recorded by the police where they involve an injury to at least one party. The main parameters of the accident itself, the involved parties and vehicles are recorded on a form by the police. These forms are collected by "Statistics Austria" and transferred into a database. This data is processed and reviewed by KfV. For the particular purpose of this study, it was necessary to add geo-referenced location data. This was achieved by calculating the geometrical centre of the municipality in which the accident happened.

3.5 *Cross-validation of Accident and Weather Databases*

The police accident database also includes information about the time and place of the accident. These values may be considered as very reliable. The question arose as to whether VERA weather information is reliable as well, or at least reliable enough to continue the exercise. VERA provides precipitation data with a 3-hours resolution, hence, either the value before or the value after the accident could be the better one. For police records, precipitation in terms of rain and PTW collisions is a somewhat Boolean variable, i.e. either rain or no rain. Virtually no collisions happen during snow or freezing rain or hail, which are the other values for precipitation in the accident recording form. Hence, it was not fully clear which category of VERA data would correspond to "rain" or "no rain" in the police data. It was found that "rain before accident" gives better values than "rain after accident" and category 1 ("very slight rain") from VERA should be considered as "no rain" in the Boolean categorization. In this case, police and VERA data on precipitation match for 94% of the cases. This was considered good enough to proceed with the analysis.

3.6 *Defining a Variable for Rain*

Here the problem is not one of identifying the weather was at the time and place of the collision. This is known from the police reports and - as shown before – it can be reliably calculated based on VERA data. The real challenge is parameterizing weather for the remaining 364 days and 21 hours of the year. This was not possible at a local level considering the calculation time and performance of the computers available and it was not useful to do this. A motorcycle rider usually does not make a decision either to ride or not based on the rain likeability at the spot of the collision he is going to have on the trip he is planning.

Hence, it was decided to calculate a variable for rain at a regional level. Austria in total has an area of 84,000 km² and consists of nine provinces, of which Vienna is the smallest (400 km²) and Lower Austria is the biggest (19,000 km²). For each of the provinces and for each day in the observation period, "rain likeability" was calculated. This calculation starts with an expert observer judgment about rain intensity at one of the 150 observatories on a scale of 0 to 5, where 0 is "no rain" and 5 is "heavy rain". From these values, within VERA a weighted average for each 16x16 km area of the grid is calculated. This is carried out eight times a day, i.e. every three hours. Using GIS data from all provinces, the values at all intersection points of the grid were summarized. As explained above, values 0 (no rain) and 1 (very slight rain) were considered as "no rain" for the new variable. All other values (2 to 5) were considered as "rain". **Rain likeability** is then a percentage of all measurements at all intersection points on the grid within one province for one day.

3.7 Rainy Days and Sunny Days

The last stage in the chain of categorization was to define an understandable, handy categorization definition for average precipitation on one day. Following several attempts, it was found that definition of a "sunny day" as having rain likeability up to 15% and a "rainy day" as having rain likeability above 75% gives the best results.

4 Results

4.1 Correlation of Weather and Collisions

Days and accidents were counted according to the procedures described above; the result is displayed in Table 1.

Table 1 Accident occurrence with respect to precipitation percentage, Austria 2002 & 2003

Percentage	Workday			Weekend			Workday & Weekend		
	Accidents	Days	Average	Accidents	Days	Average	Accidents	Days	Average
0%-15%	2943	272	10,82	1727	103	16,77	4670	375	12,45
15%-30%	879	98	8,97	338	36	9,39	1217	134	9,08
30%-45%	296	57	5,19	201	34	5,91	497	91	5,46
45%-60%	248	59	4,20	111	19	5,84	359	78	4,60
60%-75%	65	26	2,50	17	11	1,55	82	37	2,22
75%-100%	23	10	2,30	9	5	1,80	32	15	2,13
Total	**4454**	**522**	**8,53**	**2403**	**208**	**11,55**	**6857**	**730**	**9,39**

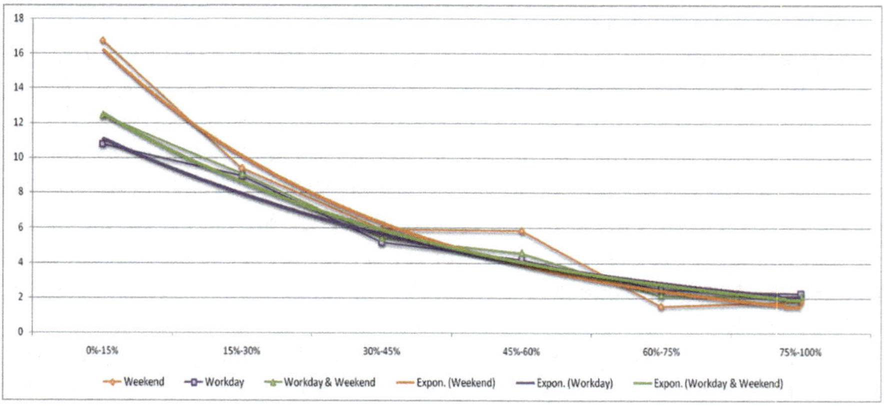

Fig. 2 Average accidents per day with regard to precipitation, Austria 2002 & 2003

As can be seen in Fig. 2, all three curves decrease virtually exponentially as the rain gets heavier. The difference between workdays and weekend days is visible from the different gradients of the respective curves. The steeper decrease of the weekend curve shows higher elasticity in the weather impact.

Table 2 shows the exponential regression functions for the occurrence of accidents per day for weather conditions in Austria in 2002 and 2003. In particular, the exponents that describe the rate of the curve's decrease and the high coefficients of determination are of interest.

Table 2 Regression curves for average accidents in 2002 & 2003 per day with regard to precipitation, Austria

Regression Curve	Function	R^2
Weekend	$y= 26{,}109e^{-0{,}474x}$	0,9061
Workday	$y= 18{,}298e^{-0{,}378x}$	0,9656
Workday & Weekend	$y= 15{,}589e^{-0{,}337x}$	0,9718

In other words, using the definition of sunny and rainy day described above:

- On sunny weekends, 8 times more motorcycle accidents happen than on rainy weekend days.
- On sunny workdays, 5 times more motorcycle accidents happen than on rainy workdays.
- On sunny days, 6 times more motorcycle accidents happen than on rainy days.

4.2 Prediction of Collision Counts Based on Weather

The main task of this study was to prove that variation in PTW collisions over the years is caused by weather, to determine the scale of the impact of weather and finally to control accident statistics for the impact of weather.

Verification of this can be made by (at least) two approaches:

It may be considered that a day with a given weather creates a typical number of accidents. The number of days of each of the rain categories within one year multiplied by the expected number of collisions (or injuries or fatalities) can then be added to gain an expected value for accidents, injuries and/or fatalities.

Table 3 shows the results of such an exercise based on the data from 2002 predicting injuries and fatalities for 2003. It can be seen in the last line that the difference between predicted and actual counts is about 3% for injuries and 11% for fatalities.

The other way is to presume a certain distribution of accident severity cases per accident for each category of rain likeability (Table 4).

Table 3 Estimated injured and fatally injured motorcyclists in 2003 on the basis of days per precipitation class

Precipitation	Days			Injured					Fatally Injured				
	2002	2003	Difference [%]	2002	Estimated in 2003	Recorded in 2003	Difference	Difference [%]	2002	Estimated in 2003	Recorded in 2003	Difference	Difference [%]
0%-15%	170	205	20,6%	2264	2730	2736	6	0%	62	75	69	-6	-8%
15%-30%	70	64	-8,6%	559	511	719	208	29%	16	15	28	13	48%
30%-45%	45	46	2,2%	268	274	257	-17	-7%	3	3	6	3	49%
45%-60%	49	29	-40,8%	252	149	121	-28	-23%	7	4	6	2	31%
60%-75%	20	17	-15,0%	63	54	18	-36	-198%	0	0	0	0	0%
75%-100%	11	4	-63,6%	31	11	2	-9	-464%	1	0	0	0	0%
Total	365	365	-	3437	3729	3853	124	3,2%	89	97	109	12	11,0%

Table 4 Estimated injured and fatally injured motorcyclists in 2003 on the basis of accidents per precipitation class

Precipitation	Accidents			Injured					Fatally Injured				
	2002	2003	Difference [%]	2002	Estimated in 2003	Recorded in 2003	Difference	Difference [%]	2002	Estimated in 2003	Recorded in 2003	Difference	Difference [%]
0%-15%	2112	2558	21,1%	2264	2742	2736	-6	0%	62	75	69	-6	-9%
15%-30%	537	680	26,6%	559	708	719	11	2%	16	20	28	8	28%
30%-45%	258	239	-7,4%	268	248	257	9	3%	3	3	6	3	54%
45%-60%	239	120	-49,8%	252	127	121	-6	-5%	7	4	6	2	41%
60%-75%	64	18	-71,9%	63	18	18	0	2%	0	0	0	0	0%
75%-100%	29	3	-89,7%	31	3	2	-1	-60%	1	0	0	0	0%
Total	3239	3618	11,7%	3437	3846	3853	7	0,2%	89	102	109	7	6,7%

It is only natural that this approach delivers more precise results, since the random variance of collisions does not enter into this assessment.

5 Conclusions, Recommendations and Outlook

This study was intended to be a pilot in order to determine whether any mathematical relationship can be defined between parameters of weather and powered two-wheeler collisions. The answer to this question, clearly, is "yes". However, the basis for this research, including only three years of motorcycle accidents from only one country, was rather small. A larger time period should be included for further investigation. Furthermore, other parameters of weather should be included with temperature as a next step. Additional research should not only look at the current situation of weather, but also at gradients, e.g. of air pressure. In particular in terms of motorcycle accidents, weather forecast information could be included.

Once the methodology has been fully developed, it should be extended to all categories of vehicles. Afterwards, accident statistics could be adjusted for the impact of weather if this is needed to analyze time series for the evaluation of particular road safety measures. If the VERA methodology is also applied in other countries, transnational comparisons could be adjusted according to the impact of weather.

In fact, any kind of phenomenon that is supposedly weather-dependent can be analyzed using the methodology proposed by this study.

References

1. Anderle, F.G., Urban, S.: Die möglichen Einflüsse von Föhn auf die Anzahl und den Verlauf von Verkehrsunfällen. Kuratorium für Verkehrssicherheit, Landesstelle Kärnten, Klagenfurt (1993)
2. Arend, H., Schwenke, K.R., Zmeck, D.: Witterungsbedingte Veränderungen der Verkehrssicherheit auf innerstädtischen Autobahnen und Hauptverkehrsstraßen. Straßenverkehrstechnik (4), 113–117 (1980)
3. Arminger, G., Bommert, K., Bonne, T.: Einfluß der Witterung auf das Unfallgeschehen im Straßenverkehr. Allgemeines statistisches Archiv. Organ der deutschen statistischen Gesellschaft 80(2), 181–206 (1996)
4. Runge in Aotomobil Revue: Wetterlagen und Unfallgefahr. Automobil Revue Nr. 34, 15. 8. 1991, 1 S. (1991)
5. Bald, S.: Untersuchungen zu Determinanten der Geschwindigkeitswahl. Bericht zum FP 8525/3 der BAST, Bereich Unfallforschung, Nr. 158, Bergisch Gladbach (1987)
6. Baumgärtner, H.: Wetter und Verkehrswesen. Zeitschrift für Verkehrssicherheit 7(1/2), 50–67 (1958)
7. Becker, F.: Der Einfluß des Wetters auf Verkehrsunfälle. Unfall- und Sicherheitsforschung Straßenverkehr (16), 90–95 (1978)
8. Behr, L.: Mehr Risiko für Wetterempfindliche? Automobil-Revue (22), 41 (1984)
9. Beine, W.: Wetterfühligkeit. Gute Fahrt im Gleichstromfeld. Auszug aus: Einfluß der Wetterempflindlichkeit auf die Kraftfahreignung. Dissertation, Univ. Graz 1975. KVDB sicher aktuell Nr. 245, p 3 (1976)
10. Biehl, B.: Zeitvariable Bedingungen der Unfallgefährdung. Übung: Psychologische Unfallforschung, Universität Mannheim Psychologisches Institut, Manuskript, Mannheim, year unknown
11. Björketun, U.: Relationships between Pavement Types and Traffic Accidents in Different Weather Conditions. VTI Topics 4(3), 3–4 (1985)
12. Brijs, T., Karlis, D., Wets, G.: Studying the effect of weather conditions on daily crash counts using a discrete time series model. Accident Analysis & Prevention 40, 1180–1190 (2008)
13. Brodsky, H., Hakkert, A.S.: Risk of a Road Accident in Rainy Weather. Accident Analysis & Prevention 20(3), 161–176 (1988)
14. Codling, P.J.: Weather and Road Accidents. In: Climatic Resources and Economic Activity, Newton, Abbot, pp. 205–222 (1974)
15. Cross, K.D., Fisher, G.: A Study of Bicycle/Motor-Vehicle Accidents: Identification of Problem Types and Countermeasure Approaches, vol. I, Final Report (September 1977)
16. Edwards, J.B.: The Relationship Between Road Accident Severity and Recorded Weather. Journal of Safety Research 29(4), 249–262 (1998)

17. Eisenberg, D.: The mixed effects of precipitation on traffic crashes. Accident Analysis & Prevention 36, 637–647 (2004)
18. Fridstrom, L., Ifver, J., Ingebrigsten, S., Kulmala, R., Krogsgard Thomasen, L.: Measuring the contribution of Randomness, Exposure, Weather, and Daylaght to the variation in Road Accident Counts. Accident Analysis & Prevention 27(1), 1–20 (1995)
19. Galvagnini, S.: Die Verkehrssicherheit der motorisierten Zweiräder. In: Diplomarbeit am Institut für Straßenbau und Verkehrswesen der TU Wien, Abteilung für Verkehrsplanung und Verkehrstechnik, Wien, vol. 29, p. 205 (Oktober 1985)
20. Hemdorff, S., Leden, L., Sakshaug, K., Salusjärvi, M., Schandersson, R.: Trafiksäkerhet och vägytans egenskaper (TOVE) (Traffic Safety and the properties of road safety). Technical Research Centre of Finland. Research Notes 1075 (1989)
21. Junk, J., Helbig, A., Krein, A.: Zum Einfluss des Wetters auf Unfallhäufigkeiten. Zeitschrift für Verkehrssicherheit 51, 143–148 (2005)
22. Johnston, P., Brooks, C., Savage, H.: Fatal and serious road crashes involving motorcyclists. Monograph 20, Road Safety Infrastructure and Surface Transport Policy Department of Infrastructure, Transport, Regional Development and Local Government (2008) ISBN 978-0-642-25570-9
23. Keay, K., Simmonds, I.: Road accidents and rainfall in a large Australian city. Accident Analysis & Prevention 38, 445–454 (2006)
24. Keay, K., Simmonds, I.: The association of rainfall and other weather variables with road traffic volume in Melbourne, Australia. Accident Analysis & Prevention 37, 109–124 (2005)
25. Leden, L., Salusjärvi, M.: Trafiksäkerhet och vägytans egenskaper (TOVE). Samband mellan beläggningens ålder och trafiksäkerheten (Traffic Safety and the properties of road safety. The relationship between pavement age and traffic safety). Technical Research Centre of Finland. Research Notes 1076 (1989)
26. Maids report, `http://www.maids-study.eu/pdf/MAIDSreport1.3_EN.pdf`
27. Ouellet, J.V., Smith, T.A., Thom, D.R., Kasantikul, V.: Environmental Contributing Factors in Thailand Motorcycle Crashes. Institut für Zweiradsicherheit (Hrsg.): Sicherheit Umwelt Zukunft IV (2002)
28. Risser, A.: Sicherheit der Arbeitswege in Wien, Kuratorium für Verkehrssicherheit (2000)
29. Saleh, P., et al.: Road Infrastructure and road safety for PTW, 2 BE SAFE Deliverable D2, Vienna (2009)
30. Sexton, B., Baughan, C., Elliott, M., Maycock, G.: The accident risk of motorcyclists. TRL Report 607, Crowthorne, Berkshire 2004, 57 S (2004)
31. Shankar, V., Mannering, F., Barfield, W.: Effect of roadway geometrics and environmental factors on rural freeway accident frequencies. Accident Analysis & Prevention 27, 371–389 (1995)
32. Sigthorsson, H., Finnsson, S.A.: Traffic accidents and weather on the airport road, Reykjanesbraut. Nordic Road & Transport Research 9(1), 14–16 (1997)
33. Steinacker, R., et al.: A Transparent Method for the Analysis and Quality Evaluation of Irregularly Distributed and Noisy Observational Data. Monthly Weather Review 2000, 128, 2303–2316, American Meteorological Society (2000)
34. Steinacker, R., et al.: A Mesoscale Data Analysis and Downscaling Method over Complex Terrain. Monthly Weather Review 134, 2758–2771, American Meteorological Society (2006)

Integrated Nowcasting System for the Central European Area: INCA-CE

Ingo Meirold-Mautner, Yong Wang, Alexander Kann, Benedikt Bica, Christine Gruber, Georg Pistotnik, and Sabine Radanovics

1 Introduction

Technological and scientific developments in nowcasting over the last 10 years have opened up new opportunities in public safety, risk management, environmental protection, and the cost effectiveness of services provided by the public and private sector. Since weather phenomena do not "stop at national state borders", the development of a truly integrated nowcasting system (one that includes the whole chain from modeling to protective action) is best achieved through transnational collaboration. Until now, no such cooperation existed in Europe and nowcasting systems were, at best, developed by single institutions. Furthermore, there was a lack of strong links between the development of nowcasting applications and the specific needs of the application side.

INCA-CE (CE for Central Europe) is a cooperation of 16 partners from 8 central European countries which bridges the gap between the development of a nowcasting system and the specific demands of the application aspect on a transnational basis. Six meteorological and hydrological institutions will work together to enhance a common nowcasting system (INCA, Integrated Nowcasting through Comprehensive Analysis, e.g. Haiden, Kann, Pistotnik, Stadlbacher, & Wittmann, 2010) as part of a transnational framework. Together with 10 partners from the scientific and application fields, the INCA-CE community will set up pilot implementations for applications in hydrology, civil protection and road safety. INCA-CE is co-financed by the Central Europe Programme of the European Union.

This collaboration began in spring 2010. Over the following pages, we will give an overview of modules already in existence, notably INCA, an outline of the project implementation and possible directions for development.

Ingo Meirold-Mautner · Yong Wang · Alexander Kann · Benedikt Bica ·
Christine Gruber · Georg Pistotnik · Sabine Radanovics
Central Institute for Meteorology and Geodynamics (ZAMG),
Hohe Warte 38, 1190 Vienna, Austria
e-mail: ingo.meirold-mautner@zamg.ac.at

J. Düh et al. (Eds.): Data and Mobility, AISC 81, pp. 107–114, 2010.
springerlink.com © Springer-Verlag Berlin Heidelberg 2010

2 INCA

2.1 Overview

The core of the pilot implementations for the application side of things, mentioned above, is now formed by the nowcasting system INCA. This is an analysis and forecasting system that is used operationally in Slovakia and Austria and which is in a test phase in several other countries (even beyond central Europe). The analytical system is based on numerical weather prediction (NWP) forecasts as a first guess and uses current observations to correct them. The forecasting takes into account the NWP error (obtained through observations) and converges to form the NWP forecast by using weighting functions. INCA provides improved numerical forecasts compared to NWP models, especially in the nowcasting range (up to +3h to +6h) but also in the very short range (up to +12h).

The INCA software is developed in several modules (essentially for each parameter) and has standardized input/output interfaces. This makes it relatively simple to install and adapt the model for use at different operating centers.

2.2 Data Sources

NWP Model Output

In the three-dimensional INCA analyses of temperature, humidity, and wind, NWP forecast fields provide the first guess as to which corrections based on observations are superimposed. In Austria, ALADIN (Aire Limitée Adaptation dynamique Développement InterNational – internationally developed dynamically adapted limited area model, developed within the ALADIN consortium led by Météo France) output is used because this is the limited area model which has been run operationally at ZAMG (ZentralAnstalt für Meteorologie und Geodynamik – Central Institute for Meteorology and Geodynamics) since 1999, and its output fields are readily available. The NWP fields are available hourly, at a resolution of 9.6 km, with 60 levels in the vertical direction. However, the INCA analysis and nowcasting methods do not depend critically on the horizontal resolution of the NWP fields and can equally be based on other NWP models, such as the LM (LokalModell – local model) or HIRLAM (HI Resolution Limited Area Model).

Surface Station Observations

The most important data sources for the INCA system are surface stations. In Austria, ZAMG operates a network of about 250 automated stations which provide measurements in one minute intervals. Data is transferred every 10 minutes (currently stations are being upgraded and in near future data will be transferred every five minutes). Additionally, around 150 hydro-meteorological stations which provide temperature and precipitation data are currently taken into account.

Radar Data

The Austrian radar network consists of five radar stations and ZAMG operationally obtains 2D radar data synthesized from the four locations, containing maximum column values in 14 intensity categories, at a time resolution of 5 minutes. Ground clutter has already been removed from the data. However, due to the mountainous character of the country, radar data is of limited use in many areas in western Austria, especially during wintertime when precipitation can originate from shallow cloud systems. Work has started on using three-dimensional radar data to obtain improved precipitation patterns for such areas.

Satellite Data

The satellite product used in INCA is 'Cloud Types' from Meteosat 2nd Generation (MSG), which consists of 17 categories. It differentiates between different cloud levels (low, medium, high) as well as different degrees of opacity. It also diagnoses whether clouds are more likely to be convective or stratiform in character.

Elevation Data

The 1-km topography used in INCA was obtained through bilinear interpolation from the global 30 second elevation dataset provided by the US Geological Survey. The resolution of 30 seconds of the original dataset corresponds to around 930 m in a latitudinal, and around 630 m (at 48°N) in a longitudinal direction.

2.3 Analyses and Forecasting System

The analyses of temperature and humidity are obtained based on the first guess of the NWP forecast and a subsequent application of the differences between forecasts and observations: NWP forecasts are interpolated tri-linearly on the INCA grid, and then differences between the model output and observations are computed to yield the correction at observation points. The corrections are spatially interpolated on the INCA grid using geometrical distance weighting.

Wind analysis is carried out in a similar manner but introduces a factor depending on topography, which translates 10 m wind to a model level wind. In the current analysis system, no additional dynamical features are introduced besides those already present in the NWP model.

INCA forecasting takes into account the local NWP error and merges this through a weighting function into the NWP forecast. A temperature forecast, for instance, consists of the observed temperature plus the temperature change predicted by the NWP model, scaled with a factor depending on the cloudiness forecast error of the NWP model.

2.4 Derived Parameters

Icing Potential

The icing potential (IP) is an empirical parameterization that expresses the degree of icing expected on structures due to contact freezing of super-cooled cloud droplets. Originally developed for icing on houses and other similar structures, this parameter could be adapted for applications in road safety where icing can lead to severe problems.

Currently, the IP is described as a function of temperature, wind speed and relative humidity. The model has been qualitatively verified and already shows promising results (see Kann, Haiden, & von der Emde, 2009).

Wind Chill

In INCA, the NOAA/NWS (National Oceanic and Atmospheric Administration/National Weather Service) formulation is used in an adapted form. In order to provide wind chill also at wind speeds below 3 mph, a linear interpolation is used. This parameter is of interest for civil protection applications.

3 Project Motivation and Implementation

There are some critical weather events which can lead to socio-economic disruptions and which clearly have a transnational character, such as large-scale windstorms (Kyrill, Emma) or large-scale flooding (e.g. August 2002). An efficient assessment of the meteorological situation in such cases is necessary to aid those agencies responsible for crisis management. The more detailed and complete this assessment can be made, the more efficiently the deployment of manpower and equipment can be organized. In order to achieve this goal, a better transnational flux of meteorological and hydrological information is vital. Weather systems and river basins do not stop at borders, and therefore they should not impede the flux of observational and forecast data either. This project proposes a new transnational framework for such information exchange which is based on a common nowcasting system and a common methodology in forecasting and warning procedures.

There are other critical weather-related events which have national or subnational character, such as a localized flash flooding or a small-scale windstorm event. Most of these are associated with thunderstorm activity. Trans-nationality is important because thunderstorm analysis and nowcasting is still an extremely challenging task and is associated with a significant predictive uncertainty. No single national weather service can claim to have solved these problems. However, all weather services in the CE partner countries have accumulated specific and different expertise on the problem over time. If combined within the framework of a common operational system, the sum of this expertise will (1) advance nowcasting capabilities in the CE as a whole, and (2) ensure a high common standard of weather information and (3) a more uniform, transnational accessibility of this information.

The inclusion of a large number of partners from the application side in the project ensures that the results obtained and recommendations given are grounded in actual real-time operational experience. By explicitly addressing the entire information chain from weather modeling through to hydrological (or other) modeling, civil protection and the road maintenance authorities, not just for a single country but all countries in the CE area, we create a sufficiently broad basis to make likely the subsequent adoption of the methodology by even more countries.

Another reason why trans-national cooperation is vital in reaching the goals of this project is that none of the involved national weather services have the technological and personnel capacity to achieve them alone. The partnership on which this project is founded reflects this necessity, but also shows the willingness of individual institutions to move beyond current national standards to create a common set of methods and procedures. While past developments in nowcasting have in fact been mostly national, this does not mean that they have been redundant or largely parallel. Rather, each country has developed specific strengths which can now be integrated into a common tool and methodology.

4 New Developments for INCA-CE

4.1 Further Developments in INCA

Within the INCA-CE project, a refinement of algorithms in INCA will be carried out. Special focus in the developments will be given to parameters which are important for applications in flood warning, civil protection (mostly storm forecasts) and road safety.

For hydrological applications, the prediction of convective precipitation needs to be improved. Currently, only a translation of convective cells is used for forecasts whereas an approach including initiation, intensification, weakening and dissipation of cells would be needed. Based on diagnostic fields already present in INCA, such as a lifted condensation level, convective available potential energy and others, the convective conditions of the area into which a convective cell is predicted to move could be assessed. Furthermore, the onset of convection is a critical factor for storm evolution and the existence of boundary layer convergence zones. Hence, such algorithms rely heavily on the quality of the wind field forecasted by the NWP.

Developments in wind nowcasting will focus on improvements in the way topographic effects (downslope windstorms, channeling) are treated, as well as in a better analysis of the downburst potential of thunderstorm cells.

For Road Safety, the nowcasting of surface temperature will be refined, especially in the temperature range close to freezing point. Clearly, all other developments mentioned above (precipitation, wind) will contribute to improved road safety applications.

4.2 Other Developments

Road safety, with regards to weather conditions, is an increasing concern of road management institutions. Beyond the icing potential computed in INCA already mentioned, an integral approach of road weather prediction is being pursued. A road weather model such as METRo, Model of the Environment and Temperature of Roads, (Crevier & Delage, 2001) will be coupled with INCA to provide detailed forecasts of the road conditions. Important forecast parameters will include road surface temperature, snow load and road condition (ranging from dry road to freezing rain). To achieve this goal,close cooperation with project partners from road maintenance organizations is necessary. In the course of this project, new measurement systems at roads will be installed and together with the road weather stations already in existence, will be integrated into the modeling chain. The improved road weather forecasts will not only contribute to improved weather warnings on roads but will also to help road maintenance organization in planning countermeasures to tackle adverse road conditions. The planning of snow clearance on roads or counteraction against freezing rain, in particular, have to be carried out efficiently.

In addition to this, other downstream applications of road safety can profit from the improved road weather forecasts. For instance, road weather information could be integrated into more technically oriented road safety concepts as developed in SAFESPOT, CVIS or COOPERS (all of which are also co-financed by the European Union).

Developments in hydrological nowcasting will focus on flash-flood forecasting and the prediction of mudslide/landslide hazards. Several partners of this project have meteorological and hydrological services combined within their institutions and thus can contribute to both the development of INCA and improvements in hydrological modeling.

5 Expected Output

The main outcome of this project is a nowcasting system that is not just state-of-the-art with respect to current meteorological and hydrological technology, but has also been specifically adapted to its day-to-day application in operational hydrology, civil protection, and road safety. It is supposed to be run operationally at all partner weather services involved in the project, and it will be tested and evaluated independently in different regions by making use of all meteorological data commonly available.

A common internet platform will be set up and will provide access to the forecasts which cover the whole CE area at a high resolution in terms of space (1 km) and time (15 min to 1hr). Due to the huge amount of data produced by the high-resolution nowcasting system (estimated at 10TB every second week), special care has to be taken regarding the concept of this internet platform. Limited data transmission rates as well as data storage capacities at individual institutions will probably not allow for the centralizing of all model outputs at one single location.

One solution could be to organize the internet platform as a WebMap server (for instance MapServer). This approach has the advantage of only transmitting data to the WebMap server as required by the user. Model output data is transmitted only corresponding to the region and zoom level as the user requested. With regards to user access to this platform, a definite scheme concerning access level has not yet been defined. The web portal is planned to be available in July 2011.

However, the output of this project will not only be the dissemination of improved short-term forecasts but also cover the specific implementation at a user level. Thus pilot implementations of complete systems also form part of the output. These pilot implementations represent complete, integrated systems along the entire information chain, from meteorological and hydrological measurement through to meteorological/hydrological models, responsible authorities, and eventually to local response and emergency teams, fire departments, road management teams, etc. There will be pilot actions for each of the three application areas of Operational Hydrology, Civil Protection, and Road Safety taking place in five of the partner countries.

Furthermore, an important outcome of this project is the actual improvement in the decision making process for end-users, brought about by the developments and activities carried out in this project. This improvement will be analyzed and described qualitatively and quantitatively. It will be summarized in the form of a comprehensive report which also includes strategies and recommendations for application of the methodologies outside the CE area. The strengths and the limits of the method, as found during the pilot actions, will be clearly described and the potential for application in fields other than operational hydrology, civil protection, and road safety, will be addressed. The report will also include feedback from individual end-users who access the weather information and warnings through the internet.

6 Conclusions

This paper gives an overview of the INCA-CE project co-financed by the European Regional Development Fund. It presents the core module, i.e. the INCA nowcasting system, which forms the basis for downstream applications in hydrology, civil protection and road safety. Further possible developments of INCA are outlined as well as developments for specific applications such as road safety. Certain core outputs are detailed to stress the trans-national character of this project.

Acknowledgements. INCA-CE is a project involving 16 partners from 8 different central European countries (see www.inca-ce.eu for details on the partnership). We would like to express our gratitude to each partner of this project. Without their commitment, this project would not be possible and this introductory article relies on the work they have carried out during the project start up phase. The INCA-CE project is implemented through the Central Europe Programme co-financed by the European Regional Development Fund.

References

1. Crevier, L.-P., Delage, Y.: METRo: A New Model for Road-Condition Forecasting in Canada. Journal of Applied Meteorology, 2026–2037 (2001)
2. Haiden, T., Kann, A., Pistotnik, G., Stadlbacher, K., Wittmann, C.: Integrated Nowcasting through Comprehensive Analysis (INCA) - System description. ZAMG report (2010)
3. Kann, A., Haiden, T., von der Emde, K.: The INCA icing potential. IWAIS XIII. Andermatt (2009)

Intelligent Transport System Architecture Different Approaches and Future Trends

Zuzana Bělinová, Petr Bureš, and Peter Jesty

Abstract. System Architecture for Intelligent Transport Systems (ITS) has been an active topic across the globe for many years. There are two basic different approaches to its creation and implementation, represented by the European ITS Framework Architecture, called FRAME, and the National ITS Architecture of the USA. Their principal differences lie in the manner and the flexibility of their use. The challenge Europe is facing is based upon the fact that there are many different states with different needs and it is therefore not possible to create a universal ITS architecture suitable for all of them. For this reason, there is framework architecture at the European level from which national or regional architectures can be created by specific states based on their individual requirements. Thus, the situation varies from state to state. On the other hand, the USA has a fixed ITS architecture, the use of which is obligatory if federal financial support for ITS deployment is desired. This paper aims to compare the two approaches and discusses the developments and ongoing activities to support ITS architecture implementation in Europe.

1 Introduction

Today's intelligent transport systems and services are based on the cooperation of subsystems interconnected by potentially different telecommunication technologies and interfaces. The development and deployment of such systems is a complex task that needs extensive preparation in order to meet the desired time schedules of implementation, high standards of operational quality and satisfied customers.

Zuzana Bělinová · Petr Bureš
Czech Technical University in Prague, Faculty of Transportation Sciences,
Konviktská 20, 110 00 Prague 1, Czech Republic
e-mail: belinova@lss.fd.cvut.cz, bures@fd.cvut.cz

Peter Jesty
Peter Jesty Consulting Ltd Warwick Lodge, Towton Tadcaster, LS24 9PB, UK
e-mail: phj@peterjesty.com

J. Düh et al. (Eds.): Data and Mobility, AISC 81, pp. 115–125, 2010.
springerlink.com © Springer-Verlag Berlin Heidelberg 2010

In other words, according to Bossom & Jesty (2009) when implementing ITS systems, complexity grows with the number of installed systems. This poses a threat to their effectiveness, manageability, maintainability, extendibility and refurbishment over time, and to overall costs. Overcoming the possible problems of having too many incompatible systems, and benefiting from the possible synergies of interoperable systems requires the use of an agreed architecture. Using an ITS architecture can save time and money during the system development and deployment thanks to the ability to analyze in advance any possible risks, different scenarios, etc. In this paper we aim to analyze these statements for the various approaches to the formalization of deployment of ITS throughout the world.

2 ITS Architectures in General

The concept of ITS architecture ranges from a relatively simple definition of a single telematics system to a “huge” definition of a complex telematics system described using several viewpoints of the system, together with its deployment plan, process and object-oriented procedures, cost benefit analysis etc. The viewpoints of an ITS architecture are constructed using different perspectives (e.g. functional, physical) and may go into different levels of detail. Beginning with the general point of view, this variety of approaches can be broken down into two basic concepts of an ITS architecture that are used in the ITS domain: high-level and low level ITS architectures (Böhm & Frötscher, 2010).

2.1 High-Level ITS Architectures

High-level ITS architectures are used to provide a description of the functionality and communications needed to supply the services expected by stakeholders from a particular ITS implementation. Within a high-level ITS architecture, the functionality is shown in the form of the "component specifications" required by ITS to implement the services, with the "communication specifications" describing how the links between the "components" will be provided. All of the "specifications" are technology independent and are intended for use as inputs to the procurement part of the overall ITS implementation process.

Making the "specifications" technology-independent gives suppliers the freedom to employ the most appropriate technical solution when tendering, while still complying with the architecture.

This type of ITS architecture is created by (or for) organizations such as national and regional governments, municipalities, research projects, etc. Aside from the FRAME Architecture, the most obvious and widely used example is the US National ITS Architecture.

Creating high-level ITS architectures is a precursor to the creation of low-level (or design) ITS architectures. The "bridge" between them is provided by the component specifications and communications specifications which are used in the procurement process.

2.2 *Low-Level ITS Architectures*

Low-level ITS architectures are used to describe the detailed design of the components and the communications, which are needed for ITS implementation. These architectures are created as part of the design activity that commences once procurement has been agreed and use the "component specifications" and "communications specifications" as their starting points. The term "components" is used because the functionality described in the specifications may be achieved using either hardware, software, or a combination of the two. The use of particular technologies will almost always form a part of the low-level architecture descriptions.

For many ITS implementations, several low-level ITS architectures may need to be developed. The actual number of these architectures will depend on how many "specifications" have been produced. In simple terms, the creation process for each low-level ITS architecture will be an expansion and refinement of the "component specifications" and "communications specifications" from which the ITS implementation is created.

3 Duality in the Development of High Level ITS Architectures

The need to have agreed ITS architectures was recognized some 20 years ago, in early 1990s.

It was motivated by the rapid research and development of services in the ITS field that needed to work together if they were to be effective. It was clear that it was not efficient to develop separate solutions for every application and that some model principle – an architecture was needed. This need was recognized on both sides of the Atlantic. Whilst the development of the Architectures was carried out separately, there is a commonality of approach. We will describe briefly the US National ITS Architecture and the FRAME (European) models of high level architecture.

3.1 *US Approach*

According to Sussman (1999) the US DOT, urged on by ITS America, triggered the development of the National ITS Architecture in 1994. The motivation for creating a National ITS Architecture was the need to describe how its various elements or subsystems would communicate, and what functions would be performed by each. A primary purpose was to describe the communications links in order to guide the development of standards, which would in turn ensure national interoperability.

A concern and reason for the development of the ITS architecture was that systems, developed freely around the country, would prevent vehicles going beyond their own regional boundaries from interacting effectively with the infrastructure in other regions. This was a particular concern for the national freight industry, which had visions of carrying 15 or more transponders in their cabs to allow interaction with various infrastructures as they travelled across the country. This was also a concern of the ITS suppliers who wished to sell hardware and software on a large, integrated, national market.

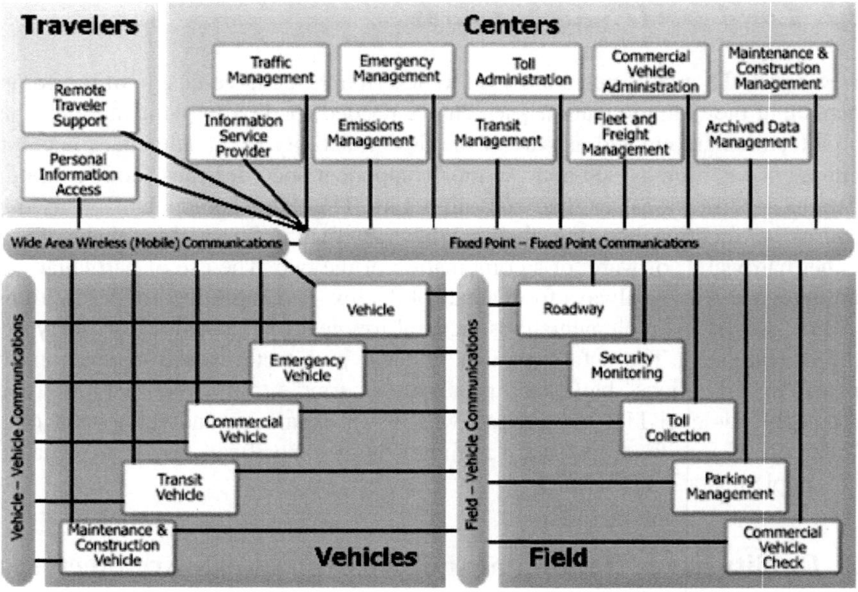

Fig. 1 Physical view of National ITS architecture (source: www.iteris.com)

The first version of the US National ITS Architecture was delivered in 1996 by a consortium under contract to U.S. DoT as a substantial and content-rich set of documents. This architecture makes a major contribution to the definition of the structure of ITS (see Figure 1). Since the architecture has to be maintained in order to stay usable, later versions of the architecture, and the tools for using it, have followed. The current version of the National ITS Architecture is found under 6.1 and is hosted at http://www.iteris.com/itsarch/

3.2 European Approach

After some initial work during Frameworks 2 and 3, in particular by the SATIN Task Force in 1996, the development of the European ITS Framework Architecture started during the 4th Framework Programme of the EC in 1998 with the KAREN project (Keystone Architecture Required for European Networks). This was developed at the request of the High Level Group in response to the need for a single reference platform in Europe, which would provide a basis for the development of ITS products and services throughout the European Union.

The basis for the creation of the architecture was the fact that, due to subsidiarity within EU member states, ITS cannot be covered by a single definitive architecture and thus a framework (meta model) of ITS architecture is needed. As it is framework architecture, its goal is to assist the development of national, regional or even project architectures.

The project produced its final ITS architecture description in a number of deliverables in 2000. Since the creation was an R&D activity, its development could have ended with the KAREN project. However, the ITS Architecture as a framework for the development of ITS systems and services was further promoted by the EC in further FP projects, and continuity was guaranteed by using the same key members of project consortium. During this time the meta-architecture started to be called FRAME (FRamework Architecture Made for Europe). Project succession in FP5 was provided by FRAME-NET and FRAME-S as support activities, and the current project E-FRAME in FP7. E-FRAME aims to include results of cooperative systems projects in the Framework Architecture, and the current version is 4. The FRAME Architecture is now managed by the FRAME Forum, which is made up of representatives from the Ministries of a number of member states.

4 National ITS Architecture

4.1 Content and Processes

The US National ITS Architecture provides a common framework for planning, defining, and integrating intelligent transportation systems. It is a mature product that reflects the contributions of a broad cross-section of the ITS community (transportation practitioners, systems engineers, system developers, technology specialists, consultants, etc.). The architecture defines:

- The functions (e.g., gathering traffic information or requesting a route) that are required for ITS.
- The physical entities or subsystems where these functions reside (e.g., the field or the vehicle).
- The information flows and data flows that connect these functions and physical subsystems together into an integrated system.

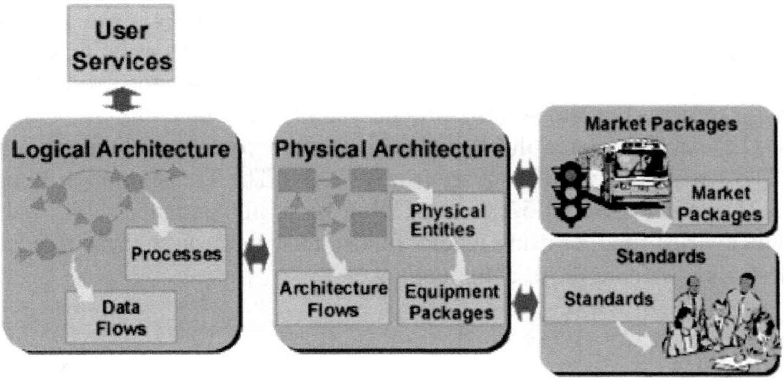

Fig. 2 National ITS Architecture view (source: www.iteris.com)

Figure 2 describes relationships between the logical architecture, physical architecture, implementation, and standards-oriented components of the US National ITS Architecture. User Services describe what the system will do from the user's perspective. A set of requirements covering each of these User Services form the basis for the US National ITS Architecture definition. The User Services entry point leads you to the full set of user service requirements and allows an easy transition between the user service requirements and the components of the architecture that satisfy these requirements.

The Logical Architecture defines the Processes (the activities or functions) that are required to satisfy the User Services. Data Flows identify the information that is shared by the Processes. The Physical Architecture forms a high-level structure around the processes and data flows in the Logical Architecture. The Physical Architecture defines the Physical Entities that make up an integrated ITS. It defines the Architecture Flows that connect the various Physical Entities into integrated systems. Equipment Packages break up the subsystems into deployment-sized pieces. Behind these entry points is the complete definition of the Physical Architecture.

Market Packages represent slices of the Physical Architecture that address specific services like surface street control. A Market Package collects together several different subsystems, equipment packages, terminators, and architecture flows that provide the desired service.

The Logical and Physical Architecture provide a starting point for ITS standards development activities by identifying the applicable architecture flows and data flows to be standardized in the US National ITS Architecture and the way in which the information is exchanged across those interfaces.

4.2 Responsibility and Applicability

The US National ITS Architecture has been designed to be used for the development of consistent "regional architectures", with the aim of facilitating the goal of national interoperability.

The organizations responsible for the "local" deployment should be willing to be consistent to ensure this goal of national interoperability, even if it is more costly to do so. In addition to this, they could see themselves as being constrained in their deployment strategies. The reason for this is the support from the federal budget for projects that are aligned with national ITS architecture.

The U.S. DOT requires the Secretary to ensure conformity with the US National ITS Architecture and Standards via TEA-21 (The Transportation Equity Act for the 21 century, Section 5206 (e)). ITS projects implemented with funds from the highway trust fund, or using federal funds, must conform to these federal rules, otherwise they will not receive the funding.

4.3 Support

The National ITS Architecture is continually supported by the government, the US DoT. Since there is a legal requirement for the use of the US National ITS

Architecture, a team maintains the architecture itself, presents workshops and seminars, checks compliance with the architecture and has developed and provides support for a software tool for the development of compliant local architectures.

The software tool which regions can use to create their own ITS architectures is called Turbo Architecture, and the current version is 4.1. Turbo Architecture is an interactive software application that assists transportation planners and system integrators, both in the public and private sectors, in the development of regional and project architectures using the US National ITS Architecture as a starting point.

5 FRAME Architecture

5.1 Content and Processes

The FRAME Architecture can be used early in the lifecycle both to convert the Stakeholder Aspirations (sometimes called User Requirements for what the ITS implementation should provide) into an ITS Architecture that satisfies them.

The first and core part of the process is the capturing of Stakeholders' Aspirations. These aspirations are simple statements about the services that end users want the ITS implementation to deliver. They can include statements about where

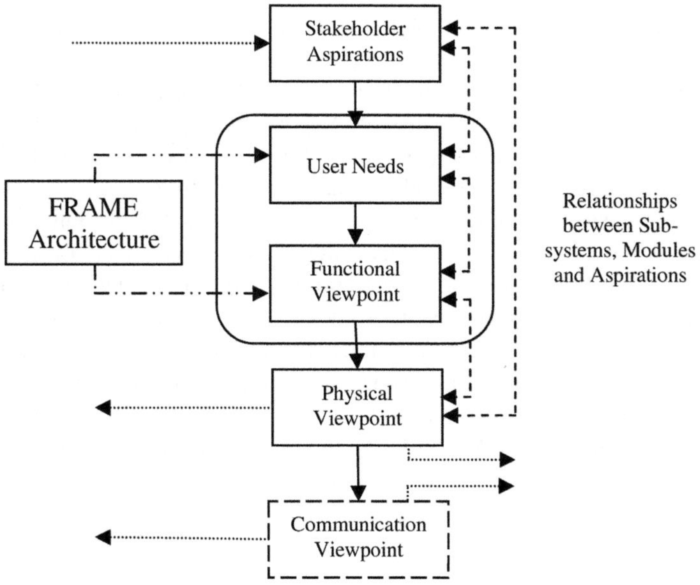

Fig. 3 Process of creating of an ITS Architecture [source: (Bossom & Jesty, 2009)]

and when the services are to be provided, and will be expressed in the stakeholders' own words. Other factors such as availability, ease of access, presentation, reliability, maintainability and, in some instances, cost can also be included in the aspirations.

These stakeholder aspirations are transformed into User Needs – from a formalized set of more than 700 needs covering most possible ITS applications and services.

The next step is to create the functional viewpoint (following the recommendations of [IEEE 1471]) – which defines the necessary functions to fulfill these User Needs. These two items (User Needs and Functional Viewpoint) are covered by the FRAME Architecture. Based on the Functional Viewpoint, but outside of the scope of FRAME, the Physical Viewpoint and subsequent Communication Viewpoint are developed.

5.2 Responsibility and Applicability

A number of European national and regional authorities now have their own national and regional ITS architectures, based on the FRAME Architecture. There have also been a number of EC funded projects that have used it.

The experiences of using the national or regional ITS architectures are mixed, and it is probably too early to draw any final conclusions. The issue is not whether the use of the ITS architecture has been beneficial – in all known cases it has been. Indeed the French Ministry reported that they were able to reduce the development and deployment time for a national speed limit enforcement system by an estimated 6 months.

There has, however, often been a reluctance to take up the use of the ITS architecture voluntarily by engineers outside the core architecture team. The reason for this is currently the subject of speculation, but the fact that the use of high-level system architectures is not normally taught at universities will be one. Fear of something new will be another. A third reason may be that many authorities are not used to making long term plans, since they have to spend all their time "fire-fighting" the latest problem, and there is no perceived immediate need to integrate the resulting systems. It is interesting to note, however, that whist the English national authority does not see a need for a national high-level ITS architecture, a number of English regions and agencies now have, or are creating, their own ITS Architectures independently of each other – all based on the FRAME Architecture.

The new European ITS Action Plan, and the ITS Directive mentions the need for ITS Architecture, though it does not go into any detail, and one can expect the situation to change as a result.

5.3 Support

The creation and maintenance of the FRAME Architecture has so far been carried out in an ad hoc manner. The EC has funded four projects (KAREN, FRAME-S,

FRAME-NET and E-FRAME), but there is no mechanism to provide long term "unlimited" support. At the end of the FRAME-S and FRAME-NET projects, the FRAME Forum was set up to provide funding from a number of Member States. The Forum still exists and a revised modus operandi is being planned to take over at the end of the E-FRAME project.

The FRAME Architecture is distributed with 2 basic tools. The first tool is called the "Browsing Tool" and is used for browsing through FRAME architecture to understand how data flows between the various functions. The second tool is called "Selection Tool" and is used for creating sub-sets of the FRAME Architecture for regional, national or project ITS architecture.

6 Comparison of EU and US Approaches

The US National ITS Architecture shows the relationship between physical components, and users choose those components that they need to satisfy their requirements. However, the European ITS Framework Architecture shows the relationship between functions only, and users first choose those that they need to satisfy their requirements, and then make their own decisions as to how they will be allocated to physical components.

The US National ITS Architecture has had continual support totaling about \$65M. Meanwhile the European ITS Framework Architecture has had intermittent support totaling around €4.5M since 1998. Whilst the European approach could be considered to be more efficient, there is still a major disparity in their support, which results in perceived differences in both their quality and their ease of use. There is also a major difference in their political backgrounds and their consequential manner of use, which is not always understood by those who stand in judgment.

The use of the US National ITS Architecture is effectively mandatory throughout the USA (unless the region does not want Federal financial support for its ITS deployments!). The Architecture comprises a set of fixed "Market Packages" and "Equipment Packages" that can be put together to fit the requirements of the stakeholders. The Market Packages and Equipment Packages are examples of "real" implementation. The European ITS Framework (FRAME) Architecture has been designed for a set of Member States for whom the concept of subsidiarity must apply. It is for this reason that everything in the FRAME Architecture is optional. This makes the use of the FRAME Architecture extremely flexible. The creator of a bespoke ITS Architecture based on this must understand it. The creation process also takes some time to complete, i.e. a number of days.

It is against this background that the US and FRAME Architectures are compared, and it is not surprising that the FRAME Architecture is sometimes criticized for being (too) difficult to use. However, since it is known that Member States undertake some ITS applications and services in different ways, the flexible nature of the FRAME Architecture will have to remain for the foreseeable future.

7 Usage of High Level ITS Architectures

7.1 Why Is It So Difficult?

In this paper, we only discuss use of high-level architectures and their impact on low-level architectures. Many people find high-level architectures very hard to understand and use. Therefore developers, who want to build the system, tend to create their own high-level architectures based on their own prior knowledge. The difficulty of using predefined, high-level architectures can be illustrated by the following statement (McQueen & McQueen, 1999):

"We have encountered many difficulties when discussing the concept of a regional ITS architecture … maybe it is such an abstract concept that most people do not see enough potential value in it to justify spending the time understanding it … the system architecture concept has been a difficult one to get people to embrace and like. It certainly seems to trouble the majority of people. Even the systems engineering community that adopted the term in system development methodology can get into quite lengthy debates about what it actually means."

7.2 Actual Usage of ITS Architecture in Europe and US

In the EU, the first two national ITS Architectures that claim some degree of compatibility with FRAME were developed by France and Italy. So far they have been used mainly for Case Studies, about which, positive reports have been received. However the French ITS Architecture (ACTIF) has recently been used as the basis for the development of a speed limit enforcement project, and for an ITS Architecture for Singapore. The Italian ITS Architecture (ARTIST) has also started to be used in the National Transport Operative Programme in projects such as ULISSE.

There are two regional and one project ITS Architecture in the UK, with two further ones being considered. Positive results have been reported from the Czech Republic and Hungary, and the FRAME Architecture has been used to some degree in many EC funded projects, in particular VIKING and COOPERS.

In the US, use of National ITS Architecture is mandatory for all projects that have federal funding.

8 Conclusion

Two approaches to the use of high level ITS architectures have been briefly described and compared. The comparison shows the strengths and weaknesses of the two discussed approaches. From our point of view, the main difficulty is to convince system developers and state officials to use it, even though the benefits are described quite clearly. Whilst the use of ITS architecture does require effort and time to fully understand and to design "any" compliant ITS system, even more work (and costs) would be incurred if an ITS Architecture were not used to achieve integrated ITS services.

The concept of ITS Architectures, as implied in the FRAME Architecture, is not well understood. They are neither used regularly by industry, nor are they covered by textbooks and taught in universities. Whilst there are a number of nations and organizations for whom their use is well understood and encouraged, there is another set who are, at best, ambivalent, and at worst refuse to consider their use – the "not invented here" principal, or believe that they go against their company's commercial interests. Part of the problem lies in the use of the term "architecture", which already has a number of seemingly different uses, and thus often invokes mental images of something else.

Acknowledgements. The authors acknowledge the financial support provided by the European Commission through the project E-FRAME (call FP7-ICT-2007-2, grant agreement number 224383).

References

1. Bělinová, Z., Bureš, P.: Study of Standards and National ITS Architectures within the E-FRAME Project. Archives of Transport System Telematics, 23–28 (February 2009)
2. Böhm, M., Frötscher, A.: E-FRAME: Deployment Issues for Cooperative systems. Deliverable D5, E-FRAME FP7 Project 224383, of ICT-2007.6.2. ICT for Cooperative Systems (May 2010)
3. Bossom, R., Jesty, P.: Using the FRAME Architecture for Planning Integrated Intelligent Transport Systems. In: ITS Prague 2009, Prague (2009)
4. IEEE 1471-2000. Recommended Practice for Architectural Description of Software-Intensive Systems
5. Iteris (August 1, 2009), Key Concepts of the National ITS Architecture. Získáno 10 (May 2010), z National ITS Architecture, `http://www.iteris.com/itsarch/index.htm`
6. Jesty, P., Bossom, R.: Using the European ITS Framework Architecture. In: ITS World Congress 2006, London (2006)
7. McQueen, B., McQueen, J.: Intelligent Transport Systems. Artech House (1999)
8. Sussman, J.M.: Regional ITS Architecture Consistency: What Should it Mean? ITS Quarterly (Fall 1999)

Looking into Detection, Model Results and Message Quality
Improved Data Verification, Liability and Quality

Andreas Schmid and Thomas Epp

Abstract. Providing publishing the "traffic situation" has become a state of the art procedure for road operators. A map showing the road network and a red-amber-green color scheme indicating the "Level of Service" or at least showing icons with traffic messages like "Jam", "Slow moving Traffic" and other incidents is a must-have.

The information provided is produced in many ways. Starting historically with manual editing work, followed by fairly sophisticated detection data processing and today using online modeling approaches and floating probe data to generate these pictures. But how can we assess how ACCURATE this information is?

This technical paper shows an approach to how the processing from detection data to event messages can be presented in a transparent way to the quality assessing engineer. At a glance, it can be seen whether the system works correctly or if any failures make the results less trustworthy. Finally, it is shown how this approach can be used to compare different sources of information and to assess their quality.

1 Today's Practical Challenges

1.1 No More Technical Service Limits

The availability of traffic engineering models and methods enable the production of "full range" services. Route planning can be offered entirely dynamically on all

Andreas Schmid
PTV AG, Stumpfstrasse 1, 76131 Karlsruhe,
Tel.: +49.721.9651-466; Mob.: +49.171.12041227
e-mail: `Andreas.Schmid@ptv.de`

Thomas Epp
PTV Austria, Planung Transport und Verkehr, Mariahilfer Str. 123, 1060 Wien,
Tel.: +43 1 599 998 005; Mob.: +43 676 721 5654
e-mail: `thomas.epp@ptvaustria.at`

J. Düh et al. (Eds.): Data and Mobility, AISC 81, pp. 127–137, 2010.
springerlink.com © Springer-Verlag Berlin Heidelberg 2010

network links taking into account the current and the future traffic a driver is likely to experience. A great number of online models and methods are available to take into account every single piece of information. Information can be of any kind such as online roadside measurements, floating vehicle data, time series providing typical traffic patterns, unforeseen incidents, planned interactions and the mobility behavior of people within a specific European region based on the local calendar.

Powerful computer processing capacities and virtually omnipresent internet technology providing high speed data links are state of the art key enablers for processing everything as required.

As a result, traffic information for now and up to 365 days in the future can be computed and updated on a virtually online basis.

1.2 Can We Understand Where a Specific Service Output Is Based Upon?

Though the promised service outputs are comprehensive, in daily operational life ITS engineers face the challenge of being able to ascertain if the generated output is good and (if a problem is known or may be assumed) that they should be able to identify weak parts of the service production and supplement less trustworthy information with better information.

The variety of methods and models and the deployment of processing capabilities needed for a "full range" online service result in rather complex architectures when analyzing system details.

This can place a significant burden on a service operator's goal to gain a good overview of the system's output quality and it can make the search to find questionable and doubtful output and the reasons behind it long, expensive and cumbersome.

1.3 Achieving Maximum Transparency

Consequently given this requirement, today's systems must be equipped with elements which support a high degree of 'insight' into relevant processing steps for the operating engineers.

2 An Example

Through the URL www.bayerninfo.de the ViB (Traffic Information Agency Bavaria – a Public-Private-Partnership organization for traffic services) provides such a "full range" service as mentioned above. Users can plan their routes or examine traffic situations. The services are inter-modal and they comprise current traffic information and short, medium and long term forecasts. This is achieved through a series of method models to compute the road traffic situation and prognosis:

2.1 *Involved Models and Methods*

ASDA/FOTO

(ASDA – Automatische Staudynamikanalyse i.e. automatic jam dynamics analysis; Foto- Forecasting of Traffic Objects) is a well approved method for assessing the precise traffic situation on detected motorway roads. Developed by Daimler Research and based on Prof. Kerners theory on three phases of traffic PTV has adopted the developments in its Traffic Platform to produce up-to-the-minute precise information where "JAM" or "SYNC FLOW" states are found on the motorway. The model can precisely capture the specific extent of these traffic situations and it is therefore used as a basis for the traffic message generation in the Siemens CONCERT system.

The ASDA/FOTO component uses the over 3,500 detectors which are designed to deliver data on volume and speed per lane, per minute, distinguishing cars from heavy goods vehicles. Further processing involves mapping the 1m resolution results from ASDA/FOTO to the abstracted road network, before the message generation produces "jam" or "heavy traffic" messages if certain thresholds are exceeded. Finally, the results are also forwarded on to the dynamic routing and mapping service through the combination of all other model results.

Short Term Forecast

The Short Term Forecast is designed to use the latest historic measurements and to identify how the determined traffic situation provided by ASDA/FOTO will evolve in the near future (next hour).

Day Forecast / Congestion Estimator

The Congestion Estimator is designed to determine potential congestion which may result from capacity reductions (e.g. lane blockages caused by road works or accidents). It provides adapted speed and flow values and congestion length information upstream to capacity bottlenecks.

Medium Term Forecast

The Medium Term Forecast is based on a classical transportation planning approach. A calibrated traffic model of Germany (VALIDATE) form the basis for an online assignment of traffic demand to the supplied road network. Both demand and supply are provided with the most actual data. Holidays and the calendar influence demand. Known road works or road closures change the 'supply'. As a feature of the assignment model approach, traffic displacement to other roads is included in the forecast. The forecast is updated every day. It provides flow and speed data on each link of the network (currently 400,000 links in Bavaria).

Long Term Forecast

Based on a transportation planning approach, PTV VALIDATE Germany provides typical day patterns of traffic for five day categories. The model is calibrated using available detector data.

2.2 Fusion of the Results

All computations have different spatial and temporal extent. Together with messages (if not already included in the models), the results must be compiled together to form a coherent picture.

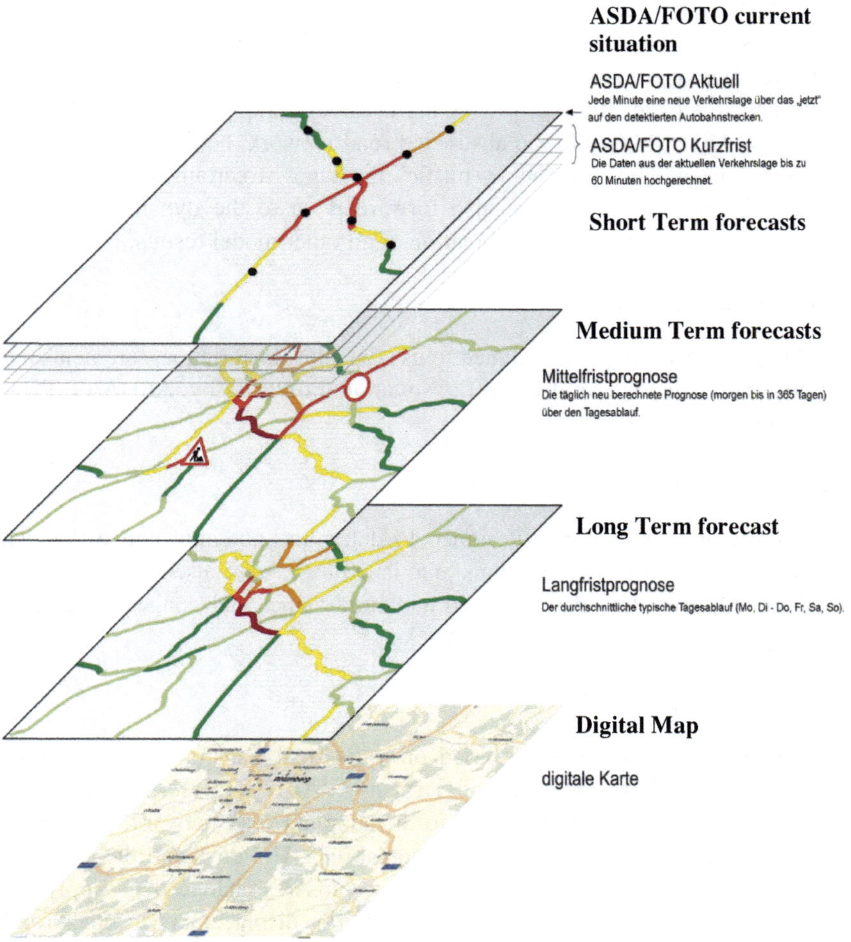

Fig. 1 Fusion of different forecast horizons and spatial extents of forecast models and methods

In the end, the service can use the composition and find an optimized path in terms of time and space for a road traveler with a minimum travel time and all the relevant information the driver will face once he or she reaches a certain section of the journey.

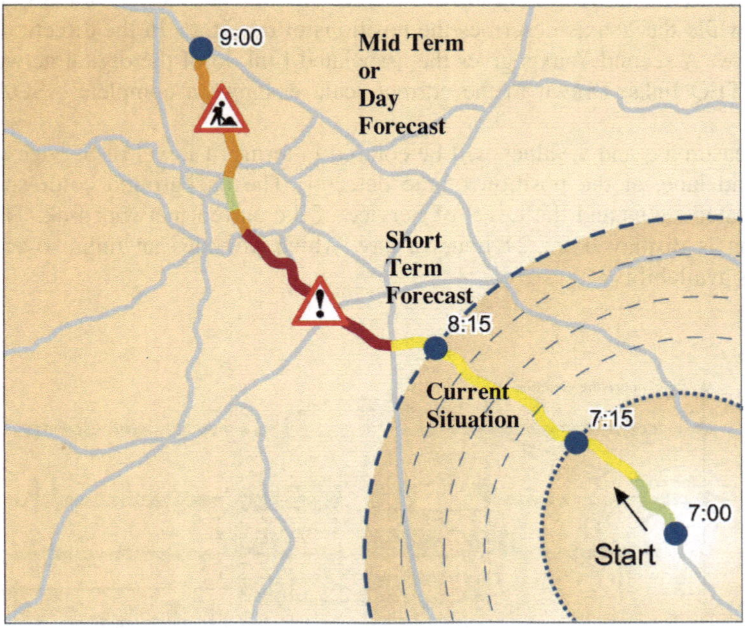

Fig. 2 appropriate usage of forecasts in routing over time and space

3 Introducing Transparency

The challenge for a service operator is to provide a "fused" image to the customer and still have means to see at one glance if anything goes wrong. A good fusion would "skew" the operator's view by eliminating suspicious data and using a more reliable fallback instead. So, a good fusion maintains service quality – but its results cannot be used by the quality survey.

3.1 ANS – ASDA/FOTO Nativ Sciagram

A successful step towards a quality survey and content failure tracking support has been made by introducing a special graphical output for the quality manager. A skilled person is able to see if the system output is consistent. Messages, Level of Service, ASDA/FOTO status and detection are composed in the form of an "X-Ray image" (or Sciagram) of a specific route on a certain day.

These diagrams are more comprehensive than the usual path-time diagrams, due to some important enhancements: the pictures show link IDs and the ASDA/FOTO route definitions, measured values for each lane and ASDA/FOTO result values for a 24 hour overview. The benefit is being able to see at a glance whether the system works correctly.

How are the diagrams organized? Values on the X-axis show minutes (1440 per day) while the Y-axis describes the positions of detectors, in the direction of traffic flow. A second Y-axis gives the associated Link Id of the digital network, e.g. NAVTEQ links. Drawn to the correct scale it shows a complete ASDA/FOTO route.

Measured q and v values will be colored in terms of Level of Service per minute and lane, at the position of the detector. The background color shows the calculated status and the Level of Service of the current network link. The information is displayed for 24 hours a day, which provides an initial overview of data's availability.

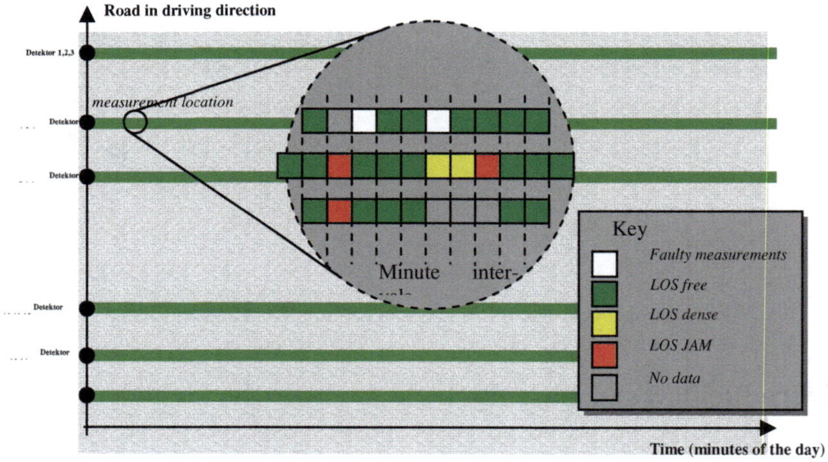

Fig. 3 ANS diagram – detector status

ASDA/FOTO results are exactly to the meter, so that the results can directly be shown in the diagram. Objects, i.e. so-called JAM- and SYNCFLOW-objects are placed according to their time (X-axis) and location (Y-axis) as computed. JAM-objects are drawn as red-colored vertical lines while SYNCFLOW-objects are yellow vertical lines.

In the final stage, traffic-messages are displayed as rectangles. They are also positioned precisely in terms of time and space, as recorded. The frame color gives an indication of the source of the message: ASDA/Foto (orange), public authorities (pink dashed), the road works information system (grey) and weather information (blue).

Diagrams are automatically created and are one of the most important pieces of evidence used for quality management. It is easy for the trained eye to check a huge amount of messages, links and motorway kilometers.

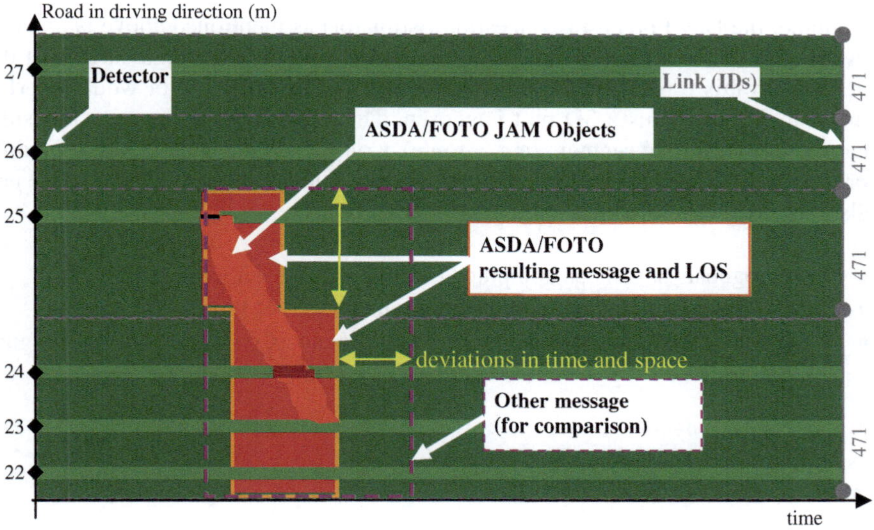

Fig. 4 ANS diagram - schema of complete view

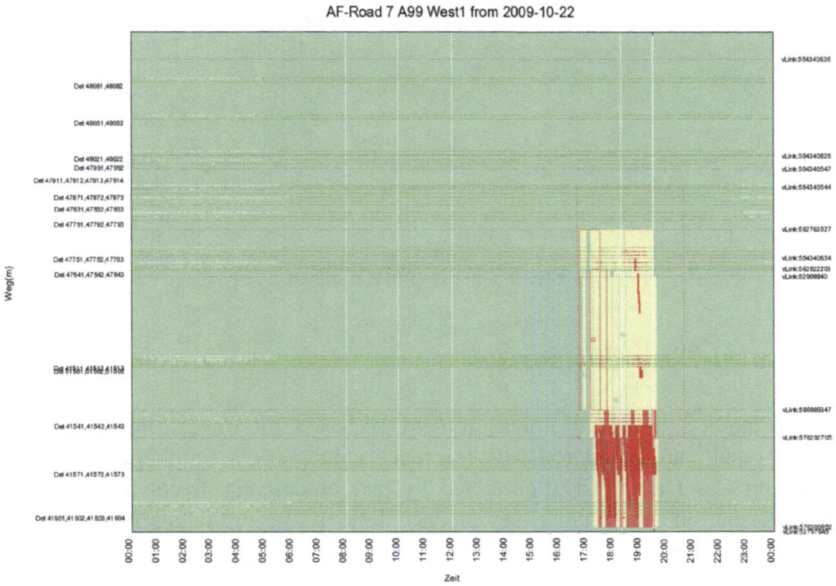

Fig. 5 real example of an automatically created ANS diagram

3.2 Prognosis Diagrams

While the current situation can be captured well in 2-dimentional diagrams, this is different for the prognosis: prognoses for a certain point in time are updated repeatedly. The closer time gets to "now" the more often updates are produced. For example, the Day Forecast (Congestion Estimator) as mentioned above is updated every 15 minutes – thus 96 prognoses are carried out for one single stretch of the network in this forecast alone. In a 2-D diagram, most of the plot would overlap on a single line. Thus a 3D plot is chosen. This plot shows all prognosis results carried out for a road segment (e.g. a route). Colors indicate which kind of module produced the result. If measurements of a current situation are available, these can also be added as a reference. At a glance it can be seen:

- If all prognosis have been computed or if there had been gaps,
- If there are significant deviations within one prognosis type, or
- If there are significant deviations between prognosis types or between prognoses and the "current situation" as actually occurred.

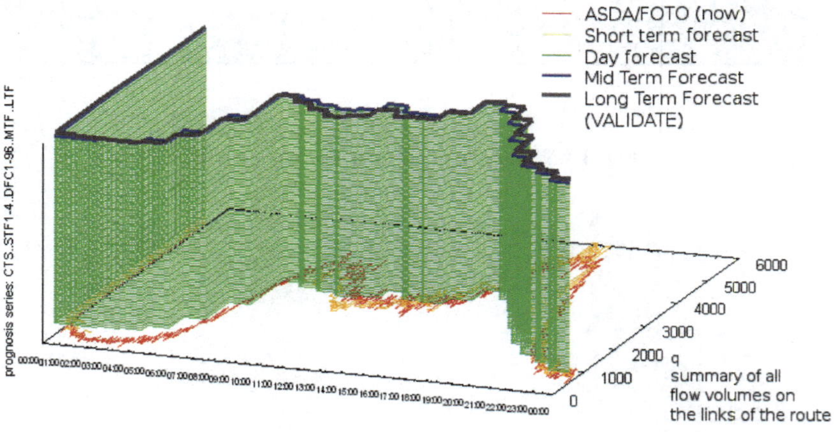

Fig. 6 3-D prognosis diagrams

3.3 Travel-Time Comparison

Another method of quality measurement is based on travel-time comparison. Forecast-results are compared with reference values. Most suitable as a reference is Floating Car Data (FCD), collected by taxi or delivery fleets. The results of FCD information are travel times at each link of the digital network, for each quarter of an hour.

One major result of the forecast-system besides the LoS is the velocity per link, which can also be easily converted into travel-times per link. For inner-city tests, major routes through the city are defined, which are divided into useful and meaningful test sections of approximately 2-3 km. These test sections are calculated every fifteen minutes.

The following picture shows the result for one route though the city. The upper part shows travel-times for both the taxi and forecast-results; times on the Y-axis are given in seconds.

The lower part shows the difference in percent. The difference is calculated using the following formula:

$$difference = 1 - \frac{forecast}{Taxi}$$

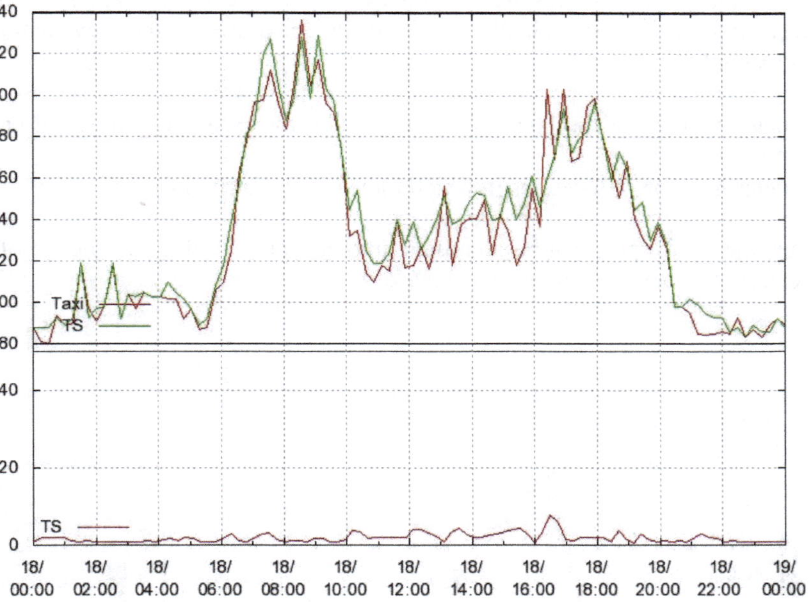

Fig. 7 travel time comparisons

Within the Vienna system (www.AnachB.at), twenty-two routes, defined on major arterial roads, are continuously and automatically tested.

4 Introducing Metrics

The transparency approaches shown in the last chapter are well suited to human interpretation and control. Even so, a higher degree of automated procedures is desirable. To allow this, metrics have to be introduced which enable automatic evaluations.

The basic assumption of the introduced metrics is that a comparison between two values is made. One value represents the assumed "true" situation (e.g. a measured speed or flow which occurred) and the other value represents the model calculation (e.g. a forecast). A quality index is calculated, looking at the difference between the two values. Current tests use the following approach to calculate the quality index:

$$Quality \quad Index = \frac{\left| computed \quad value - "truth" \right|}{("truth")^{n}}$$

In the approach the n is determined based on an empiric study.

The metric provides results for all network links (i.e. all locations) and all time steps which can be statistically evaluated. Practical uses of these evaluations are for example:

- determination of the hours of the day when forecasts are quite good or fairly poor
- determination of locations in which forecasts are fairly poor

Thus the weak spots can be identified automatically. With a knowledge of these weak spots, counter measures can be taken.

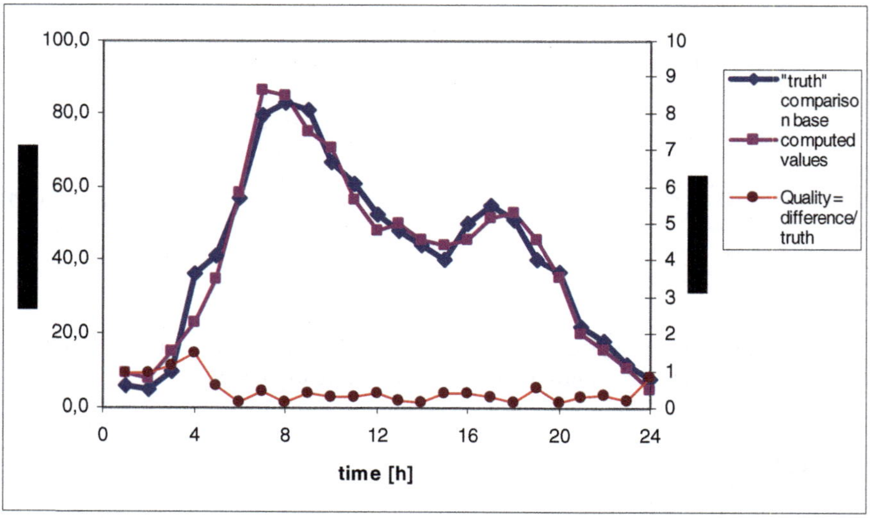

Fig. 8 Quality Index metric example for a series of measurements

5 Summary

ITS systems can be developed for delivering comprehensive "full range" services. In order to ensure the quality of the services, the steps prior to the data fusion through to end result must be made transparent for assessment. Automatically generated diagrams in 2D and 3D offer insight to these system process results. In addition, the introduction of metrics supports automatic evaluations in order to identify weak spots.

Today such approaches are not well established yet. The examples shown provide an insight into practices which have been developed in projects using the PTV Traffic Platform.

The approaches shall inspire future applications of ITS products. Transparency and quality measures support not only improvements such as failure search in terms of good engineering work, they are also very valuable in terms of convincing customers, that ITS really can deliver good, usable and trustworthy output.

References

1. Kerner, B.S.: The Physics of Traffic – Empirical Freeway Pattern Features, Engineering Applications and Theory. Springer, Berlin (2004),
Website with online traffic information – jam and heavy traffic messages and level of service color are based on ASDA/FOTO where motorways have sufficient detection, http://www.bayerninfo.de

Maintenance Decision Support System (MDSS) ASFINAG / Austria

Experience of a Comprehensive Winter Maintenance Management System

Werner Seidl and Thorsten Cypra

Abstract. Some of the major objectives in road operation are safety, environmental protection, economics and the necessary optimization of these issues in delivering quality winter maintenance services. In order to increase road safety and optimize traffic flow, high quality prediction and sensor technologies are required as well as appropriate winter service treatments at the right time. Given the complexity of meteorological, traffic and winter service processes, winter services require a comprehensive Winter Maintenance Management System.

ASFINAG (Autobahnen- und Schnellstraßen-Finanzierungs-Aktiengesellschaft) plans, finances, maintains and tolls the entire Austrian motorway and highway network covering 2,178 kilometers. This figure includes over 300 kilometers per direction in tunnels and some 300 kilometers per direction over bridges. ASFINAG is wholly owned by the Austrian Federal Government.

In order to benefit from synergies and to reach common, high quality standards – particularly in winter maintenance – ASFINAG has decided to launch a specific, company-wide project: the MDSS ASFINAG.

The MDSS system is a common, robust, modular and scalable platform able to support the decisions of all the maintenance depots within the two service companies of ASFINAG. Boschung Mecatronic AG has been selected to develop and to provide the MDSS system.

Werner Seidl (Dipl.-Ing)
ASFINAG Maut Service GmbH, Am Europlatz 1,
A-1120 Vienna, Austria
e-mail: werner.seidl@asfinag.at

Thorsten Cypra (Dr.-Ing)
Boschung Mecatronic AG, Rte d'Englisberg 21,
CH-1763 Granges-Paccot, Switzerland
e-mail: thorsten.cypra@boschung.com

J. Düh et al. (Eds.): Data and Mobility, AISC 81, pp. 139–150, 2010.
springerlink.com © Springer-Verlag Berlin Heidelberg 2010

The objectives of the MDSS system are Road Weather Information System (RWIS), data concentration, the integration of weather forecasts and combination of RWIS and forecast data as well as the integration of operational data to devise situation maps for comprehensive decision-making support.

The MDSS system brings together all of the RWIS stations distributed over the road network. This provides a common interface for all users, who will have the opportunity to access data from neighboring regions.
The Austrian road network has been divided in about 230 Road Weather Segments corresponding to micro/meso-climatic regions. Specific weather forecasts for up to 72 hours will be established for each of these segments.

One of the main benefits of the MDSS system will be to combine local RWIS station data with weather forecast data to develop the road surface and road weather information of the corresponding Road Weather Segment so as to aid decision-making.

1 Introduction

Mobility is an important location factor in economies; uninterrupted traffic is a basic requirement for the effective development of the economy and society. A significant increase in traffic volumes on Austrian motorways has an effect on the quality of traffic flow. Statistics from the Austrian Federal Ministry for Transport, Innovation and Technology BMVIT shows, for example, an increase of 70% in the transportation of goods over the Austrian Alps between 1994 and 2004 (BMVIT, 2006). The aim of road management must be that a road user can drive a certain distance in a predictable time as reliably in winter as in summer. As a consequence, the standards of maintenance on motorways have become very high. Therefore, winter maintenance in particular must be organized in an optimal manner and turnouts need to be implemented extremely quickly.

This means that those in charge of winter maintenance need a comprehensive Winter Maintenance Management System, because of the complexity of the meteorological, traffic and winter service processes. A comprehensive Maintenance Decision Support System clearly displays all of the important elements in one view, such as the weather forecast and road condition information, local RWIS station data, location of winter service vehicles and operation data etc. for real-time, future, and past events. The combination of local measurements (RWIS-Stations) and weather forecasts allowing detailed predictions and early alerts for each forecasted Road Weather Segment is especially useful.

2 Winter Service Operation on Motorways in Austria

ASFINAG was established in 1982 and is wholly owned by the Austrian Federal Government. A contract signed in 1997 between the Federal Government and ASFINAG gave the company additional powers and responsibilities: by virtue of this contract, ASFINAG holds usufruct rights related to land and facilities belonging to the primary federal road network and owned by the Federal Government and has the right to collect tolls and/or charges from those who use this land and facilities.

2.1 The Network

ASFINAG plans, finances, maintains and tolls the entire Austrian motorway and highway network covering 2,178 kilometers. This figure includes over 300 kilometers per direction in tunnels and some 300 kilometers per direction over bridges.

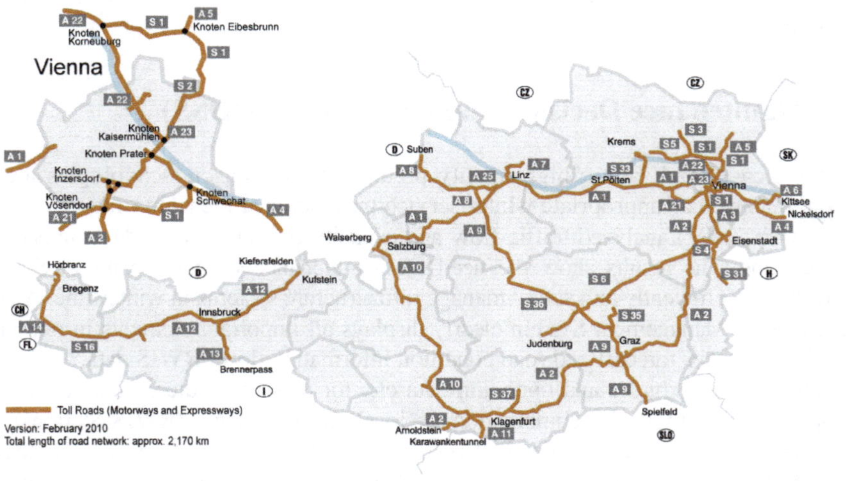

Fig. 1 ASFINAG road network (motorways and highways)

2.2 Organization of Winter Service

ASFINAG Service GmbH (SG) and ASFINAG Alpenstraßen GmbH (ASG) are responsible for ensuring the operation of the motorways and highways including the maintenance and servicing of the necessary infrastructure and provision of optimal customer information. Within their respective service region (ASG for Tyrol and Vorarlberg, SG for the other provinces), they are responsible for all activities relating to operational maintenance, maintaining electrical equipment, as well as asset and traffic management. All structural and operational activities are geared towards meeting strategic targets such as increasing customer satisfaction, ensuring network availability, optimizing performance standards and economic efficiency.

Some 2,178 kilometers must be maintained, and rest areas and parking facilities cleaned. In the winter, snow has to be cleared while renewal and repair works start in spring; green areas have to be tended, trees and bushes cut and technical installations inspected. Tunnel facilities also have to be washed and safety installations inspected. Staff are based at the tunnel control centers all year round to monitor events in the tunnels closely and must be able to intervene immediately if an incident occurs thus ensuring road safety. Staff at the ASFINAG Traffic Control Centre in Inzersdorf keep a constant watch over what is occurring over the entire road network. On their screens, the ASFINAG operators can see all sections equipped with traffic control systems and can react to any disruptions in the traffic flow. Thus road users are informed quickly of any obstructions and ambulance, fire brigade, police or motorway operation and maintenance facility staff can quickly be called to the scene. The winter service team, for instance, consists of some 1,500 staff on duty 24 hours a day to ensure driveability on motorways and highways. To this end ASFINAG uses state-of-the art technical equipment and tools ranging from precise weather forecasts to computer-controlled gritting.

3 Maintenance Decision Support System (MDSS)

To increase road safety, high quality prediction and sensor technologies are required as well as appropriate winter service treatments at the right time.

Cypra (2007) analyzed traffic flow and capacity on highways under wintery road conditions. The results show the need for Maintenance Decision Support Systems (MDSS) to efficiently and safely manage infrastructure systems in wintertime. A comprehensive Management System clearly displays all important elements in one view, such as weather forecast and road condition information, local RWIS data, location of winter service vehicles and operation data etc. for real-time, future, and past events. Especially useful is the combination of local measurements (RWIS-Stations) and weather forecasts allowing detailed predictions and rapid alerts for each forecasted Road Weather Segment (road section with similar micro/mesoclimatic conditions). The collection, combination and visualization of all this information for the Maintenance Decision Support System gives the person in charge the possibility of making decisions efficiently and of managing dynamically, which increases road safety and improves traffic flow.

3.1 Structure and Functions of MDSS

The MDSS is a common, robust, modular redundant and scalable platform able to support the decisions of all of the maintenance depots within the two service companies of ASFINAG. Boschung Mecatronic AG has been selected to develop and provide the MDSS system for the ASFINAG network.

This MDSS management system is a web-based network solution with numerous functions to support the management of road network. It is available internally or externally over a standard web browser. The servers of the MDSS system are located at two locations inside of an internal network with secured ports to the internet.

This MDSS system is a comprehensive winter maintenance management system for monitoring the current and expected road conditions, controlling deployments and automatically recording and displaying all the data and reports. The new features of this software are, on the one hand, the intelligent linking of weather information and periodic road condition data from RWIS stations, taking into account local general conditions and providing road condition predictions not just for local points but for whole roads sections, so-called Road Weather Segments. On the other hand, a management system has been created through which the user receives a large volume of information presented clearly within situation maps, e.g. RWIS station data, visualized road condition prognoses, RWIS reports, location and activities of winter service vehicles, on one user desktop.

The following overview shows the functional principle of the MDSS system, together with the key input parameters and dataflow, which the system uses to prepare the appropriate prognoses, centralized databases and visualized information. The data exchange between the different sources and the import into the management system is ensured by defined interfaces.

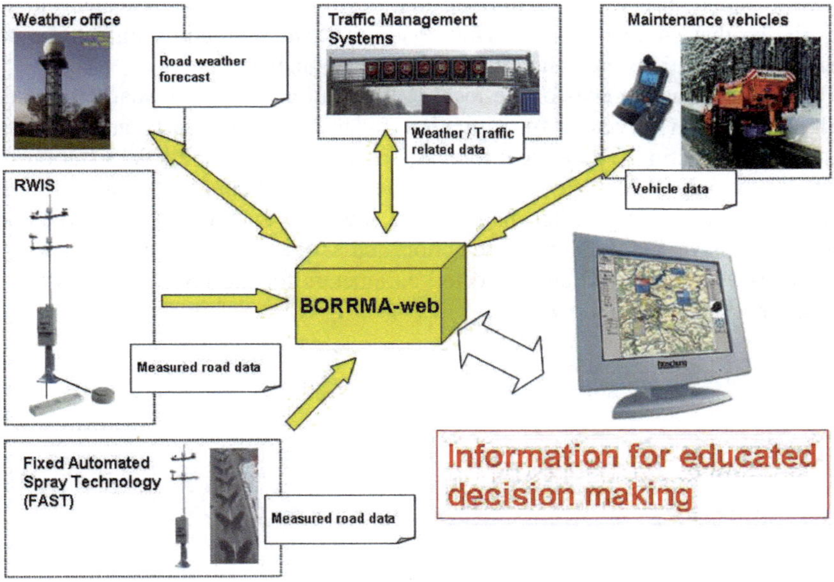

Fig. 2 Functional principle of the Management System (BORRMA: Boschung Road and Runway Management System)

Besides the integration of station data from the present RWIS and FAST (Fixed Automated Spray Technology), the MDSS system will provide a direct multivendor-capable means of receiving data from measurement stations via the new Measurement Station SOAP Interface (MSSI). SOAP stands for Simple Object Access Protocol, a general network protocol for exchanging data.

In addition to the road weather forecasts and the RWIS station data, the data and predictions of the RWIS stations are of high importance in launching and controlling winter maintenance activities. On the situation map, the winter maintenance manager can, for example, see all of the RWIS stations as flags, and is informed of

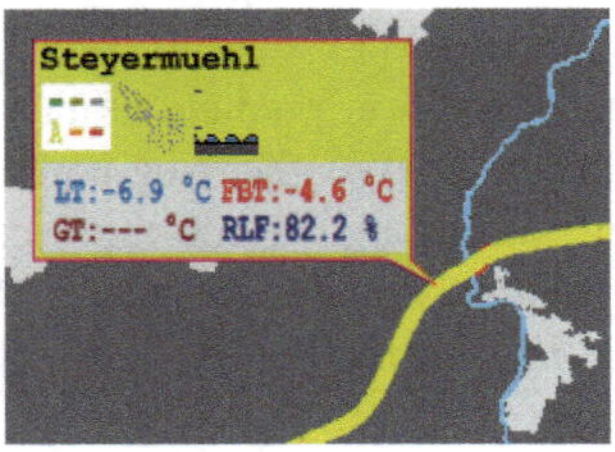

Fig. 3 RWIS station with individual configured parameters and colored station alert level 1

expected local dangers by a station alert. ASFINAG has colored the alerts in a common way (green = no alert, yellow = lowest alert until red = highest alert).

Several steps must be followed for a route-related road condition process. The network under observation is divided into sections according to meteorological criteria, so that similar microclimatic conditions can be found within these segments. These sections are called Road Weather Segments.

Using the measurements of the local RWIS stations, detailed weather forecasts and precipitation forecasts (precipitation radar), local forecasts are calculated based on nowcasting. These local forecasts usually relate to the RWIS station site. This nowcasting is calculated for at least the next two hours and is updated regularly. When controlling winter maintenance activities, it is these two - three hours that are decisive for the correct selection and scope of winter maintenance deployments. The measured and expected parameters for the RWIS stations can be viewed in a separate window.

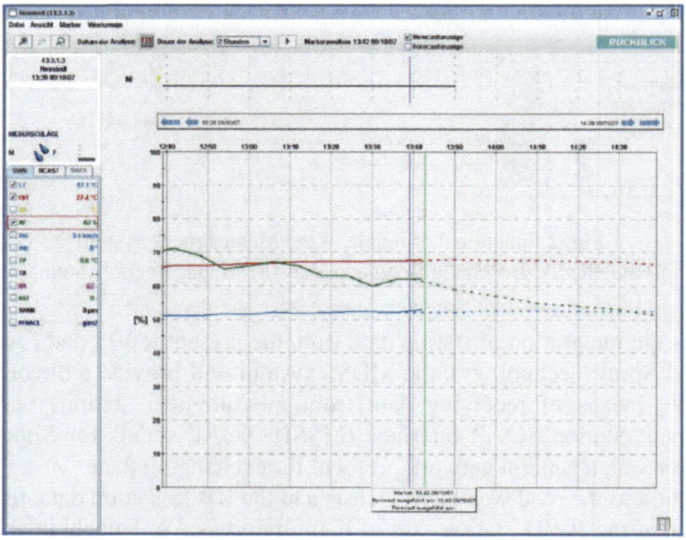

Fig. 4 Measured and nowcast data for an RWIS station

Using the local prognosis, the forecast function is used to calculate a long term (three-day), route related prediction. This forecast is based on the results of the nowcasting, long term weather forecasts and additional general conditions. Risk levels are determined for time periods for each road weather segment using innovative decision-making methods. These risk levels are then displayed on the dynamic map for each road weather segment according to time. Each time a road weather segment is clicked, the individual risk levels (no risk, risk level 1 to 3) are displayed at the bottom right of the user desktop along with the time of the occurrence. This gives the road maintenance manager an overview of the current status of their entire road network and, using a time slider, of the forecast road surface states. Besides the road surface conditions, it is also possible to display the air temperature or road surface temperature for all the road weather segments on the map. The principle of nowcasting / forecasting is shown in the figure below.

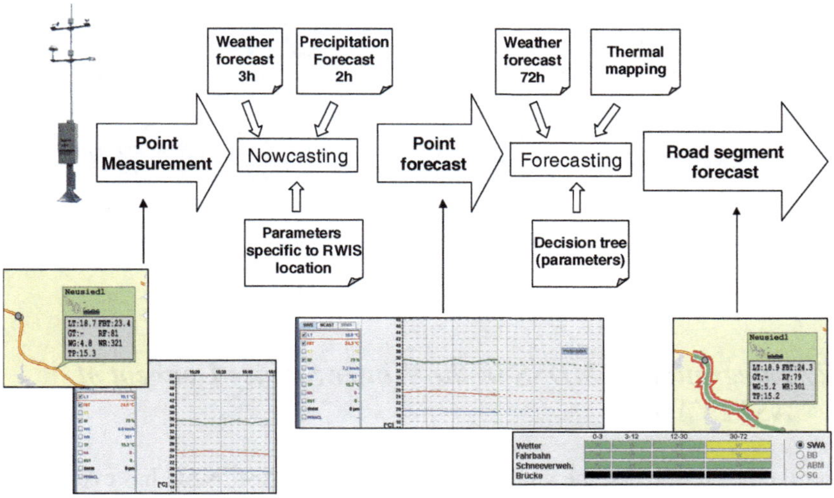

Fig. 5 Principle of nowcasting – forecasting of MDSS system

This gives the winter maintenance managers a tool with which they can control the winter maintenance deployments efficiently and plan the optimum personnel and vehicle requirements over a longer period of time.

In addition to road condition predictions, the actual measures of winter service are important informational indicators for the decision makers. The Management System of ASFINAG also shows the locations and measurements of the winter service vehicles in real time. This allows the decision makers to be informed about the actual situation and to manage the winter service in a dynamic manner.

Furthermore, automatic data recording of the winter maintenance deployments as part of vehicle management can be used to create complete documentation without any manual effort. The data in the database can be evaluated in order to fulfill road safety obligations, invoice contractors, create balances and statistics which, amongst other things, considerably reduces the amount of administrative work. This is even more important in order to meet the growing requirements for fast, functional and economic winter maintenance as part of a comprehensive winter maintenance management system.

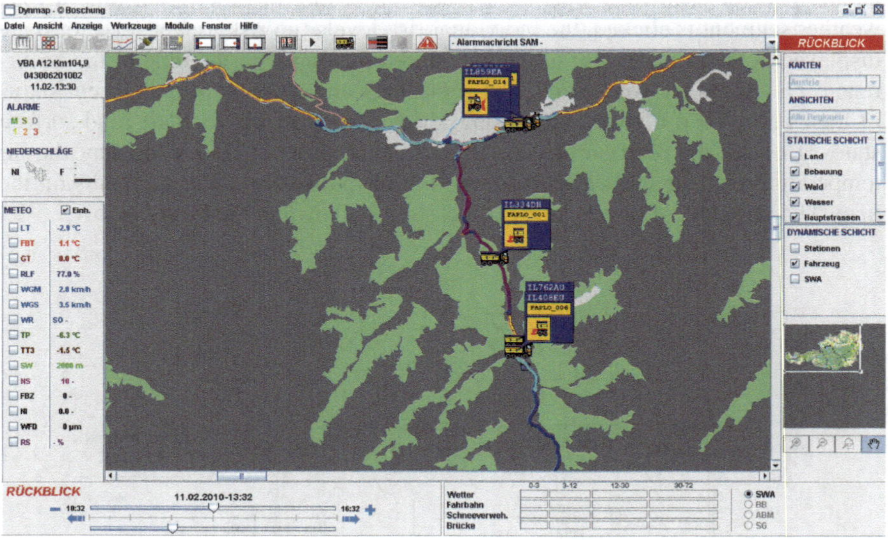

Fig. 6 Vehicle and RWIS data on MDSS screen

3.2 *Integration of MDSS in the Winter Service Concept of ASFINAG*

Over the last few years, the winter service concept of ASFINAG has evolved from the use of weather forecast reports received once a day by fax, local measurement data from ice warning / RWIS systems with stations at specific sites on the road network as well as manual recording of interventions, to the use of a fully integrated solution with a set of specific features. The MDSS represents this integrated next-generation tool focusing on winter maintenance services and is one cornerstone of meeting the strategic targets for road operations such as increasing customer satisfaction, ensuring network availability, optimizing performance standards and economic efficiency.

The MDSS solution contributes to the ability of the service organization to adapt and develop winter service processes to meet these targets. The solution enables us to follow the ongoing organizational changes e.g. performing a winter service surveillance role for a larger service area at tunnel control centers to support the winter service teams. Furthermore, the solution provides specific demand-oriented views for all user groups involved in winter service to support the daily business of truck drivers, those in charge of deployments as well as the road masters and operators at the surveillance centre.

To provide accurate weather forecast information to the winter service organization and to the ASFINAG Traffic Control Centre in Inzersdorf, ASFINAG developed a company-wide Road Weather Forecast Solution. This System delivers weather forecast information for the next 72 hours for the whole road network for each Road Weather Segment, e.g. weather forecast data and text messages, weather situation maps accompanied by satellite and radar information. This Road Weather Forecast System will be added to the MDSS to represent all of the information needed by the user within one portal. MDSS will provide "integrated" early alerts according to a common alerting scheme by integrating the specific alerts from the Road Weather Forecast System.

As an additional feature, images of the ASFINAG camera network will be available in MDSS in the near future to optimize the assessment of the situation. At this stage of integration MDSS can support a decentralized service organization as well as centralized service organization.

4 Experience of MDSS in Austria

The development of the MDSS solution was conducted in several steps with close cooperation in particular with winter service professionals at the maintenance depots. One of the main benefits of the solution is a demand-oriented presentation of all the information needed by the user group within one view. Therefore integration with the existing systems (RWIS systems, TMS system, Road Weather Forecast Service) and the securing of consistent data views between the different systems were crucial to customer acceptance. The integration of the service vehicles into the system generates a next level of benefit by supporting the organization with a comprehensive situation map including the current maintenance activities to manage the deployment phase, automatically generated reports of winter service activities and recorded information directly accessible from the user-client for returning to the real situation that happened to ensure the specific improvement of winter maintenance measures, deployments and strategies within the winter service teams.

Appropriate and resilient predictions for the future weather and road condition events are of great relevance to the decision-makers in particular for the next two / three hours but also for the time outside of regular working hours, especially evening-night-morning periods and weekends. This requirement focused specifically on the development of the Road Weather Forecast System, in particular within the design of the user interface.

Besides the road weather forecast, MDSS delivers in particular predictions for road weather / road condition scenarios, risk levels and early warnings for the next 72 hours according to a decision tree. The decision tree was designed at ASFINAG using a classification scheme for weather and road conditions reflecting relevant situations for winter service operations and a decision engine assessing the type of event and level of risk. The calculation is based on the results of the nowcasting procedure for a time period of the next two/three hours. The type of event and level of risk is displayed on demand for each Road Weather Segment in the situation map.

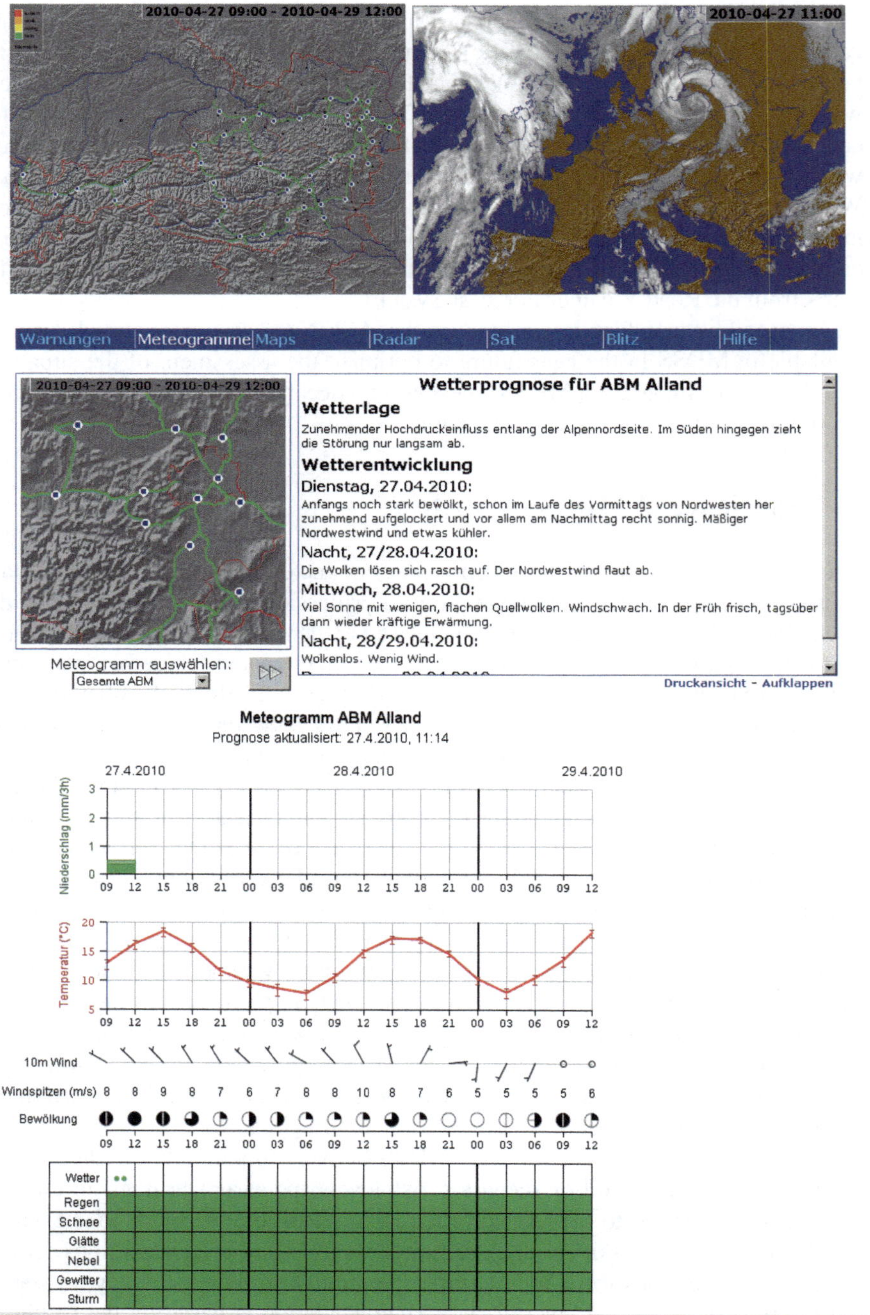

Fig. 7 Road Weather Forecast – Examples of Views

The prediction procedure was deployed for 2 maintenance depots with diverse climatic situations last winter for operative testing and basic calibration. The results were basically good and issues for further consideration were identified, e.g. improvements in the decision tree for specific weather situations, further development of the module to provide a list of messages for weather / road condition events, optimization of the presentation of alerts to the user and further simplification of user interaction.

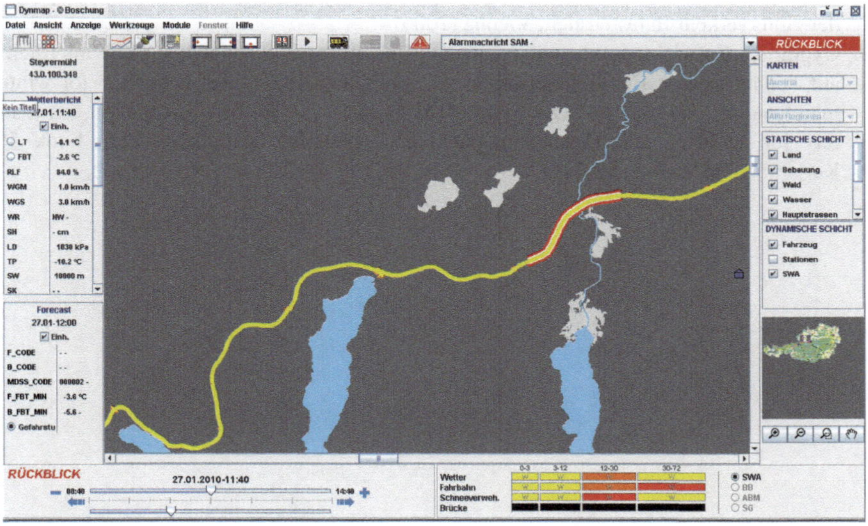

Fig. 8 MDSS situation map with risk level for the selected Road Weather Segment

Another important issue was how to derive information for reporting and analysis. The MDSS provides some online reports to record event situations in the short term. For more sophisticated reporting and analysis, resilient MDSS data will be forwarded to the business data warehouse.

The system provides a high degree of flexibility in order to adapt e.g. classification schemes, the decision tree and user signalization at different user levels to provide predictions clearly related to the actual set of measures for optimized road condition treatment. The system is expected to play a key role in enabling further optimization of winter maintenance service activities.

5 Conclusions

Complex decision-making situations, such as those which occur in winter maintenance management, require supporting systems to detect dangerous road conditions and winter events early enough to plan and control road maintenance properly. A differentiated knowledge about what is happening in the near future based on the road conditions and winter service resources is necessary to organize an efficient, fast winter service. In the field of winter maintenance, this Maintenance Decision Support System (MDSS) is a comprehensive tool to monitor the weather and road conditions and control, manage and log the winter maintenance measures.

References

1. Bundesministerium für Verkehr, Innovation und Technologie (BMVIT), Alpenquerender Güterverkehr in Österreich, Vienna (2006)
2. Cypra, T.: Entwicklung einer Entscheidungsmethode für Maßnahmen im Winterdienst auf hochbelasteten Bundesautobahnen (Development of a decision method for winter maintenance measures on heavily travelled motorways), Veröffentlichungen des Instituts für Straßen- und Eisenbahnwesen der Universität Karlsruhe (TH), book 56, Karlsruhe (Germany) (2007)

METIS
Road Weather Monitoring and Presentation System

Tomáš Pospíšek and David Konečny

Abstract. METIS is a unique single-platform and technology-independent solution for sharing and presenting road weather data. It integrates all available weather information sources, e.g. road weather stations, remote sensing products, traffic cameras, weather forecasts etc., and presents the information using maps, animations, images, graphs, tables and text with emphasis on the present and future weather situation on the roads. METIS is complex yet user-friendly software that gives a comprehensive overview of the weather in real time. METIS is designed in particular for winter road maintenance professionals to provide all the information needed for correct and timely decision-making.

1 Introduction

The main objective of the METIS presentation and monitoring system is to offer well-researched information about the current conditions on the roads for the winter road maintenance professionals as well as offer an outlook into the near future so that those responsible can make a valid decision while managing the maintenance personnel and resources.

There is a significant need for professional road weather information systems enabling proactive winter maintenance [1]. The information for winter road dispatchers comes from many sources and the situation may sometimes be confusing. METIS focuses on the transformation of the data into understandable information and its clear presentation.

METIS is universal and adjustable software for the integration and presentation of all kinds of available road-related weather information sources (e.g. road weather stations, precipitation radar, satellite images, forecasts etc.). METIS is not

Tomáš Pospíšek · David Konečny
CROSS Zlín, s.r.o., Louky 397, 763 02 Zlín,
Czech Republic
e-mail: www.cross.cz, pospisek@cross.cz, konecny@cross.cz

J. Düh et al. (Eds.): Data and Mobility, AISC 81, pp. 151–161, 2010.
springerlink.com © Springer-Verlag Berlin Heidelberg 2010

directly connected to a specific manufacturer of road weather station or provider of weather data and can be implemented independently of the type of technologies used in the case of open data formats.

The Meteo Poll Server runs in the background of METIS to collect all the input data from weather stations and third-party FTP servers. To connect to individual stations, the Meteo Poll Server uses mostly cable modems (land line), GSM/GPRS and CGU servers. Meteo Poll Server processes the data so they all have uniform structure and are centrally stored and shared by METIS.

METIS is a fully web-based online application currently developed for the Microsoft Windows platform by using the PHP and Delphi programming instruments. Some functionality is implemented using the JavaScript scripting language and Ajax techniques.

2 Installation and Operation

Before using METIS on the user's PC, the user has to apply for a new user account by completing a simple online registration form. Once we have processed the user's data and prepared a new user account, a confirmation e-mail is sent with instructions for installing and launching the application.

Since METIS is distributed as a single executable file, there is no need to install the application on a local PC. The METIS Viewer will start immediately after launching the EXE file. METIS Viewer serves as a framework for user authentication and application security.

The METIS application can be used on any modern computer which is connected to the internet and has Microsoft Windows operating system and Internet Explorer (6.0 or higher version) installed.

We intend to make METIS more open so even **non-Windows users** can access METIS. This winter, we will launch a new METIS version, which will be accessible using a regular internet browser.

Users can switch between different languages in the "Language" menu of the application. English is the standard. Other languages can be easily incorporated on demand.

3 Application Functional Description

Basically, METIS is a common road weather information system with all ordinary components [2] such as status map, data from road weather stations and weather forecasts. But what makes it unique is that it incorporates several special products / functions and its control is very simple and intuitive.

3.1 Status Map

The status map is a central element of the RWIS. In summary, it gives an overview of the current weather situation across the entire monitored area. The map application consists of 4 basic structural elements: Main map, Toolbar, Layers and key, and Reference map.

The main map in the central part of the application contains the map, including a graphical and numerical scale and scale slider. The toolbar above the map contains a search module and action buttons for movement in the map, map reload and map help (see Fig. 1.).

As default, the status map is set with a hidden right panel, which contains a list of the available data layers, a map key and a reference overview map. The right panel is displayed or hidden by pressing the relevant button on the toolbar.

The data available within the METIS map application is divided into three categories: meteorological, background and external data. The **meteorological data** represent measurements of road weather stations and radar (precipitation) and satellite (cloudiness) images. The **background data** serves for orientation across the territory and contains towns, road network and administrative borders. The **external data** represents the data connected in the application from other sources, such as orthophoto maps connected via WMS. The individual layers can be turned on and off using the check marks in the layers' list.

The status map displays point localized **road weather stations**. Based on different construction and equipment, we distinguish the types of stations identified on the map with a particular symbol. The filling color of the symbol refers to the status of the current station warning. **Station warning** is one of the most valuable

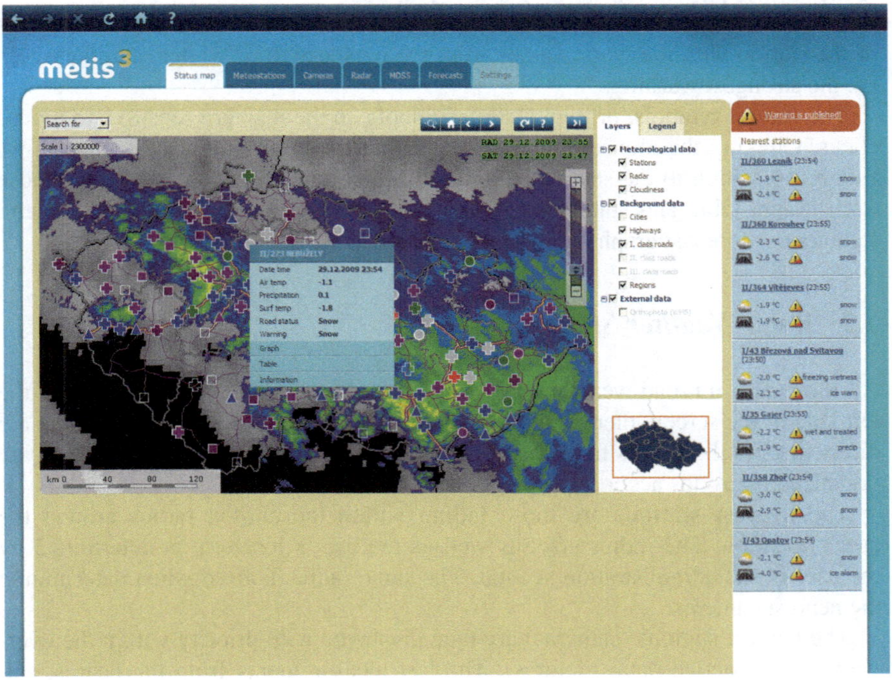

Fig. 1 METIS status map

characteristics of the system, which is why it is also shown in the status map. It is a key added value to the basic measured data. It is a classified value based on the so-called slipperiness conditions [3, 4], via analysis of the current measurements, and warns the user against e.g. snow, frost or ice. From the following winter season, the map will enable the display of other measured characteristics from the road weather stations.

Besides the basic stations shapes, the map also contains another symbol for the **joint stations**. This symbol is a reaction to the overcrowding of the map with station symbols (currently there are approx. 350 stations in the Czech Republic). The new symbol virtually connects the nearest stations with one representative symbol. This merging is based on analysis of station closeness, depending on the current map scale. This prevents any undesirable overlapping of stations, and ensures their precise geographical localization while the map remains well-organized. The new symbol acts as a regular station and can be used to access the measurements of stations merged within it.

The map application is **interactive**. The user is allowed to move freely in the map field, as well as zooming in or out. Hovering the cursor above the active elements in the map displays brief, descriptive information.

Extensive information is available when accessing the road weather station, or joint stations. Descriptive weather station information always contains updated data on measurements. As well as the station name and time mark of the record, data regarding air temperature, precipitation intensity, temperature and condition of the surface and station warning is also featured.

The descriptive information for the joint station symbol always states the values of the station with the worst values from the perspective of winter maintenance – i.e. the strongest warning.

The station symbols, or joint station symbols, in the map are sensitive to mouse clicks. Clicking the mouse button and selecting the item within the menu allows a direct transition to the selected tab of the Meteostations page. Users can choose from graph, table and station information. Depending on the station equipment, a camera or forecast graph may also be available.

3.2 Road Weather Stations

All the data from road weather stations is located under Meteostations tab. On the left side of the screen, there is a menu with a list of all the stations. First, there are direct links to the details from the nearest road weather stations. Other stations are organized in groups according to regions.

The **nearest stations** are those falling within the chosen radius around the user's location. The radius size, as well as the user's location, is determined by the user in the administration section. The same radius is used when determining the nearest cameras.

The nearest stations (cameras) are then displayed with priority within the overviews of the actual data (pictures). The last measurements from the nearest stations are also displayed on each page of the METIS application by the right edge of the screen, in the 'bonus' panel.

After switching to the Meteostations page using the main tab, a **table overview** of the nearest meteostations is located in the central part of the screen. The overview of the nearest stations is switched on as default. The user has the option of displaying the overview of the actual measurements for the regional or all meteostations. Where the current data is not available, the entire station row is grayed out.

3.2.1 Detail of a Road Weather Station

Clicking on the station will bring the user to the station detail page. Users can click on the station in the overview table, in the menu on the left-hand side, in the bonus panel on the right-hand side or go to station details through the status map or MDSS forecast map.

The station detail page contains five tabs – Graph, Table, Camera, Forecast graph and Information – according to station equipment. The controls located in the upper part can be used to move to a detail of another station or to browse through the measurement history. The stations are organized in a cycle based on their geographical location – not alphabetically. This order better reflects the users' logic and is, in addition, partially dependent on the selected tab. If the Camera tab is active, pressing the arrow to move to the next station will display a station equipped with a camera – i.e. the stations without cameras are skipped.

Station graph is one of the most frequently used elements of the METIS system. It displays the maximum values measured by the road weather station (see Fig. 2). The *line chart* displays all of the measured temperatures, precipitation intensity, and, in some cases, other special measurements. The scales of the axes are always adapted to display the data properly. The horizontal axis corresponds to time.

A *banded chart* is located under the line chart, and shares its time axis. The banded chart displays the status values which cannot be displayed numerically in the line chart, i.e. precipitation type, road condition, station warning and possibly warning presented on VMS, which can be controlled by a road weather station.

The measurement graph is designated as 24-hours. The graph of the actual values always shows the last 24 hours. The graph of historical values always shows the entire night-time period, with midnight in the middle. Such a display is more practical for the dispatchers of the winter maintenance service, given the more demanding maintenance carried out during the night period.

The station graph is **interactive**. Particular values can be obtained by clicking on the graph. Values are shown at the right side of the graph. Transition to the next measurement can be carried out by clicking on the graph or using the arrows on the right of the graph.

The station table displays any available station measurements, in numerical or text form, for the last 24 hours in the current display, or for the selected 24 hours in the case of displaying historical data. Each row represents one measurement. The data is sorted from the most current on the top to the oldest on the bottom. The values are colored according to a specific color code, referring to the value meaning.

Fig. 2 METIS station graph

Pictures from cameras at the relevant meteostation are available under the **Camera** tab. There are a few options for displaying the historical pictures. The arrows on the right, above the main picture, can be used to move forward or backwards, and the scroll down menu displays the times of all the available pictures in the currently displayed day.

The user can also browse through the individual pictures by using **'picture previews'** under the main picture. The available previews are 1 and 2 hours older and, in the case of historical pictures, 1 and 2 hours more recent than the currently displayed picture. Clicking on these previews displays the relevant picture in full size.

The **forecast graph** is one of the METIS items which provide the user with an outlook into the future. The forecast graph is available only for the selected weather stations with the appropriate technology, or within the reach of a specific forecast module, which may be for example, MDSS, the winter road maintenance decision support system made by CROSS.

Besides the expected development of the values, the graph gradually also displays the measured values, so it is possible to monitor the level of accuracy of the individual forecasts. The measured values are differentiated from the expected values in the line chart.

The **Information** tab provides any and all available information regarding the relevant road weather station. For example, the station type, overview of its equipment, detail of its location, photographs, reason for the any possible outage, etc.

3.3 Cameras

The Cameras tab contains any and all available pictures from the cameras located at the road weather stations or from traffic cameras. In the case of availability, it is possible to integrate output from video cameras.

The page layout is similar to the Meteostations tab. A menu on the left-hand side with links to the nearest cameras in front with all the cameras sorted into groups according to regions and an **overview of the latest pictures** in the central part of the screen with the following sorting: Nearest, Regional and All.

For practical reasons, the pictures are organized in the overview according to geographical location, i.e. not alphabetically. This means that the cameras from one region are close to one other.

Clicking the selected camera will redirect the user to the camera detail page. A main picture is located in the central part of the page and **previews** of previous and, possibly, the next pictures with a time distance of around 1 and 2 hours from the currently displayed picture are shown under the main picture. In addition, there are arrows and a scroll down menu above the main picture for transition between the individual pictures for the selected day. The controls for transition to the next camera and browsing history remain the same as for the meteostations.

3.4 Remote Sensing Products

The Radar section gives the user access to a set of distance mapping products. These are mainly satellite pictures of cloud and precipitation areas from the land meteorological radars. Other products of a nation-wide or global character can also be included in this section, such as, for example, output from the numeric weather forecast models.

The left-hand side of the page contains a menu for selection of the required product. Incorporation of the individual products of the remote Earth research always depends on actual availability at the place of implementation of the METIS application. In the Czech Republic, for example, the **forecast radar** is available. This is a special product of the CHMI, showing the expected development of the actual precipitation field for 90 minutes into the future.

The page contains the real-time imagery, as well as historical pictures accessible via the so called 'interactive radar'. The real-time imagery always contains the pre-prepared animations for the last few pictures of the individual data sets, without the option to adjust the animation in any way.

The **interactive radar** allows the display of pictures for the selected time horizons in the picture history. At the same time, it allows the combination of more data sources, and displays a multi-layer animation. In addition, the user can choose various background layers important for orientation in the area and set the parameters of the animation display (animation speed). The course of the animation of the interactive radar can be controlled using the control buttons.

3.5 Forecasts, Including MDSS

The Forecasts tab provides the users with text forecasts issued by the national weather service [5] and other special forecast products. Typically, it is a special set of forecasts prepared according to the needs of winter road maintenance.

Text forecasts are divided into regional and general. The regional forecasts are always prepared for individual regions or for significant roads. The general forecasts have an impact on the entire monitored area and are of a more general nature.

The weather forecasts for winter maintenance are divided, according to the forecast horizon, into short-term and medium-term. Precise forecasts for a horizon of a few hours are the most important for winter maintenance. However, for planning the dispatcher service and scheduling personnel, forecasts for the next few days to come are also important.

3.5.1 MDSS

MDSS is the *winter road maintenance decision support system* made by CROSS for dispatchers of road maintenance. MDSS is the road forecast module that serves to provide **a precise short-term local forecast of road slipperiness** in winter. Road condition, road surface temperature, freezing point temperature and snow volume are forecasted for the upcoming 24 hours with 1-hour interval **for each 1 km segment** of the road network [6].

With a simple end-user interface (see Fig. 3) and field-proven, precise predictions, MDSS is a highly unique prediction system for road professionals.

3.6 SMS / E-mail Warning System

The SMS / E-mail warning system is an optional structural part of METIS system. The user has the option of receiving notifications via short SMS messages or e-mails in the case of upcoming dangerous situations which are present or are predicted to occur in the user's surroundings.

Setup of the warning system is performed by the user in the administration section and can be adjusted at any time when using the application. The warnings are generated in two cases. The first case is when the warning of the relevant phenomenon occurs in a text forecast from the national weather service. The second type of warning is activated based on measurements of the selected road weather stations.

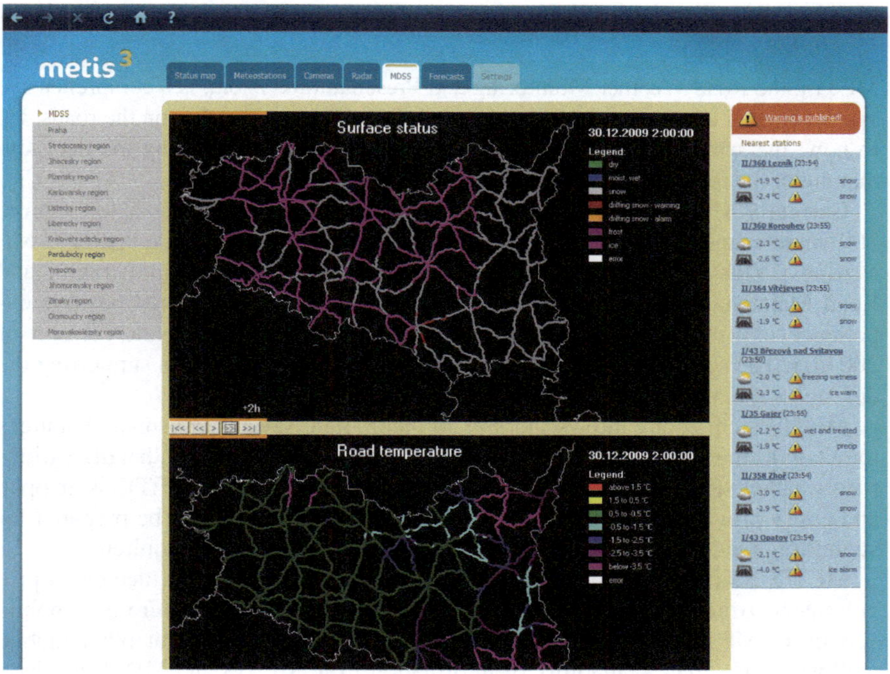

Fig. 3 MDSS forecast map

4 Key Features and Benefits

Below we briefly present several reasons why METIS really is a unique platform for winter road maintenance, making full-featured information support possible.

- *Single software platform* for sharing and presenting all kinds of available road-related weather data. The maximum quantity of relevant information from different sources is concentrated into single application.
- *Real technology independence and open architecture.* METIS can incorporate data from the road weather stations of all producers in the case of open data formats and other diverse data sources.
- *Interactive data presentation* using maps, animations, images, graphs, tables and text.
- *Complex, yet user-friendly application.* METIS has been designed with special care, concentrating on the data required by maintenance professionals but made very user-friendly so even new users can navigate around the application very quickly.
- *Web-based online access* using thin client. The web-based architecture enables presentation of the most up-to-date information to users.
- Single executable file, no installation required on local PC
- SMS / E-mail warning system

5 Conclusions

METIS, the Road Weather Monitoring and Presentation System, is a comprehensive system created to present all of the collected data related to weather on the roads. Up to now, the system has proved itself to be fully suited to supporting the decision-making of the responsible maintenance authority personnel.

Currently METIS is operating in the Czech Republic and Serbia for approx. 300 maintenance professionals from some 15 maintenance organizations. Mr. Marian CVRKAL (Winter Maintenance Operations Chief, Administration and Road Maintenance of the Pardubice Region) argues: "We have had positive experience of working with METIS on a daily basis since the winter season of 2004/05. The system has proved itself to be a very valuable tool in supporting the decisions of our operators."

METIS architecture allows the user to incorporate several information sources to obtain as much relevant information as possible. The system can handle various road weather stations, making it technology-independent. Since METIS is an open and easily programmable system, different specific interfaces can be prepared for submitting the METIS data to superior traffic-related systems, if required.

METIS is constantly being updated and upgraded by a team of dedicated programmers. Among the newest features to be launched in near future is a mobile version of METIS for the increasingly popular iPhone platform or other mobile platforms and direct availability of information from METIS and MDSS for drivers using RDS-TMC technology [7].

References

1. Boon, C.B., Cluett, C.: Road Weather Information Systems: Enabling Proactive Maintenance Practices in Washington State (2002), `http://citeseerx.ist.psu.edu/viewdoc/download?doi=10.1.1.129.3538&rep=rep1&type=pdf` (accessed May 10, 2010)
2. Standing International Road Weather Commission. RWIS web guide: A Guide to Road Weather Systems (online), SIRWEC, `http://sirwec.org/documents/rwis_web_guide.pdf` (accessed May 7, 2010)
3. Sulan, J.: Hoar-frost as a Danger Phenomenon for the Road Traffic. Meteorol. Bull. 59(2), 37–42 (2006)
4. Norrman, J.: Slipperiness on roads – an expert system classification. Meteorol. Appl. 7(01), 27–36 (2000)
5. Sulan, J., Škuthan, M.: Road Meteorology in Operation of the Czech Hydrometeorological Institute. Meteorol. Bull. 58(2), 33–40 (2005)
6. Pospíšek, T., Balajka, M., Konečný, D.: Maintenance Decision Support System. In: ITS Prague 2009, Conference Proceedings, Prague, March 30-April 1 (2009)
7. Burtwell, M.: Improvements to snow and ice control on European highways. Meteorol. Appl. 8(04), 475–479 (2001)

Abbreviations

CHMI	Czech Hydrometeorological Institute
MDSS	Maintenance Decision Support System made by CROSS Zlín
VMS	Variable Message Sign
RWIS	Road Weather Information System
WMS	Web Map Service

Naturalistic Driving

A New Method of Data Collection

Martin Winkelbauer, Anita Eichhorn,
Fridulv Sagberg, and Agathe Backer-Grøndahl

Abstract. In a typical naturalistic driving study, the subjects' own cars are equipped with devices that, over a longer period of time, continuously monitor various aspects of their driving behavior in an unobtrusive way and without the presence of a test supervisor. This includes aspects of vehicle movement, driver behavior, and the direct environment. Naturalistic observations of pedestrians and cyclists can be carried out using site-based fixed cameras.

Naturalistic observations provide information that is difficult or even impossible to obtain through current research methods. For example, analyses of crash statistics or in-depth crash investigations cannot provide much information about behavioral issues preceding a crash or about near misses. Observations by means of instrumented vehicles or simulators do not encourage the test subjects to behave in a normal (naturalistic) way, since they are generally well aware of the experimental conditions.

Experiences in the US have indicated that the naturalistic approach may give a reliable picture of the driver's normal behavior and makes it possible to observe and analyze the interrelationship between the driver, vehicle, road and other road users under normal conditions, in conflict situations and in actual collisions.

The PROLOGUE project aims to assess the feasibility and usefulness of a large-scale European naturalistic driving study and to establish a scientific and organizational basis for this new type of research.

Martin Winkelbauer · Anita Eichhorn
KfV Austrian Road Safety Board, Austria,

Martin Winkelbauer
Kuratorium für Verkehrssicherheit (KfV),
Schleiergasse 18, 1100 Vienna
Tel.: +43 (0)5 77 0 77-1214
e-mail: martin.winkelbauer@kfv.at

Fridulv Sagberg · Agathe Backer-Grøndahl
TØI Institute of Transport Economics, Norway

J. Düh et al. (Eds.): Data and Mobility, AISC 81, pp. 163–176, 2010.
springerlink.com © Springer-Verlag Berlin Heidelberg 2010

The work in PROLOGUE comprises a number of small-scale pilot studies in different research areas, including novice drivers, in-vehicle information systems, and vulnerable road users, through site-based observations. The pilots are preceded by an inventory of the technical, methodological and organizational issues. The work has identified potential areas of application and research questions for which the naturalistic approach would have an added value. Ongoing naturalistic studies have been reviewed and summarized. A survey has been carried out to support the preparation of a catalogue of applications and research topics for future naturalistic driving studies.

1 Introduction

Research into the field of traffic safety mainly focuses on collisions. Most of the research work tries to analyze quantitative and qualitative data on collisions to learn about accident causation and to develop countermeasures. There are several different methods of collecting data about accidents, such as police records, reviews of court files or on-the-spot accident investigations.

30 years ago, the "Wiener Fahrprobe" (Viennese driving test) was developed as a means of driver observation for the purpose of driver assessment. The method was quite sophisticated, including a comprehensive observation program. However, the observation period was approximately one hour and there were two observers in the vehicle, which does not fully motivate a driver to behave in a normal manner.

Instrumented vehicles have been used in the past, mainly for field operational tests (FOTs) or similar experiments. In most of the cases, the vehicles are filled with cameras, computers, wires and plugs. In many cases, an observer is on board as well; at least one assistant is needed to start the observation device and to save the data after a test drive. Frequently, such observations have been carried out on standardized routes and the test subjects were driving just for the purpose of the test.

Simulator studies offer opportunities to simplify research work, but still the time spent in the simulator, i.e. the amount of information, is limited, and not all questions can be reliably answered in a simulator.

Nevertheless, using instrumented vehicles and simulators is a step towards observing "normal behavior" instead of analyzing accidents. There are weaknesses in all these methods, which are soon to be overcome with the use of a new method.

The PROLOGUE (PROmoting real Life Observations for Gaining Understanding of road user behavior in Europe) project aims to assess the feasibility and usefulness of a large-scale European naturalistic driving study. PROLOGUE is funded by the European Commission within the framework of FP7. It started in August 2009 and will finish in July 2011. There are nine partners from five EU-countries, Norway and Israel. PROLOGUE is coordinated by SWOV, the Road Safety Research Institute of the Netherlands. The project will deal with all aspects of ND studies, starting with a literature survey and defining potential research questions. PROLOGUE develops proposals on aspects of data collection, i.e. which parameters to measure and which technology (sensors, data storage) to use. There will be a focus on how to analyze the enormous amounts of data that will be collected through naturalistic driving research. And PROLOGUE will conduct five small-scale pilot studies using different approaches (in-vehicle and site-based observation), different technology for data collection and different methodologies for data assessment and analysis. Legal and ethical issues will receive particular attention. All this work will feed into a set of recommendations to establish a scientific and organizational basis for a large-scale European Naturalistic Driving experiment.

2 What Is "Naturalistic Driving"?

Naturalistic Driving (ND) aims to collect data about drivers as they act in everyday "driving life": going to work in the morning, driving during working hours, e.g. for business trips, driving for work, going home in the evening, going out at night and leisure excursions at the weekend. The observation of the driver and data relating to the vehicles shall be collected as unobtrusively as possible, i.e. without continuously reminding the driver that they are a test subject in a scientific study. The behavior of the driver shall be as close as possible to the behavior he/she would exhibit without being a test subject.

"It has been proposed that naturalistic studies can be classified as one of three basic types of study; 1) baseline/normative/exposure studies where the aim is to investigate driving behavior and performance per se, 2) critical incident/near-crash studies where the main aim is to characterize and investigate such incidents and 3) system-focused studies where the aim is to study the driver's interaction with in-vehicle systems (Llaneras, Freedman et al. 1999). This is a useful categorization, even though studies of type 2 should include actual crashes as well as critical incidents and near-crashes, and investigation of behavior leading to these events should be of main interest" (PROLOGUE D1.1, Backer-Grøndahl et al. (2010).

3 Naturalistic Driving Observational Data

ND uses a data logger. It is a task of the PROLOGUE project to define what data should be collected by a European ND driving study. It is quite clear that there are two conflicting goals of data collection: on one hand, it is our basic interest to know as much as possible. However, collecting many parameters also means a lot of data must be stored and then analyzed afterwards. Typically, ND also includes video recording with at least one video camera at the front of the vehicle. There may also be cameras to the rear and to the side of the driver as well as at the drivers' feet. In particular, cameras raise issues of data protection and privacy for the driver and the same goes for recording an audio signal from the interior of the vehicle. Videos are most intensive in terms of disc space. This makes it clear that - on the other hand – the comprehensive recording of many parameters, both in terms of quantity of parameters and resolution, limits the duration of an experiment or creates problems in terms of data transfer from the vehicle to a central storage and the total amount of data.

A typical large scale ND experiment involves 10,000 drivers and an observation period of about 6 months. It is estimated that this produces data of approximately one terabyte. This challenging assumption comes from the fact that in the US, after having carried out a 100-car-study, this forms the framework data for the "10,000 car study" that is currently being carried out. The PROLOGUE project aims to adopt such framework conditions in Europe, i.e. to define the necessary extent of a large scale European ND study.

It is important to differentiate between the core activities of ND experiments. In practice, ND studies consist of two main parts: data collection and data processing. However, this is more or less the same as the procedures for accident data collection. As an example: recording, processing and providing accident data is a continuous activity which, in itself, is not scientific work. It is just a necessary routine task before the results, i.e. an accident database, can be used for scientific purposes. The same goes for ND: in principle, data collection, i.e. carrying out a large scale ND survey, is not a scientific task; it is simply preparation for scientific work to be carried out afterwards. Nevertheless, just as it is the case for accident databases, ND requires a great deal of scientific input and work in preparing and updating the methodology.

This preparation, for accident databases, has been carried out 27 times in the 27 Member States of the European Union, creating 27 different procedures. All the problems that appear today of compiling these 27 national databases together into a European database (CARE) should be avoided for a large scale European ND study. Hence, the scientists who will analyze ND data at a later date are now tasked with setting up the ND data collection in a way that later facilitates research work in the best possible uniform way.

4 What Can Be Measured by Naturalistic Driving?

This question is quite similar to the previous one and addresses the technological aspect of data collection within ND studies. There is a huge variety of information that may be interesting to investigate; there are also a huge variety of different sensors to collect this information. However, disc space and processing capabilities are limited and a careful choice has to be made before starting a large-scale experiment. PROLOGUE addresses these issues in its work package 2. Besides video recording, decisions have to be made on whether to store or not, and if so, at which resolution:

- speed
- acceleration
- lateral (lane) positioning
- eye tracking
- position relative to other vehicles
- event identification aids
- environmental factors
- road type (map matching)

- physiological data of the driver
- CAN data (i.e. all kinds of information from the vehicle)
- etc

In total, PROLOGUE's deliverable D2.1 (Welch et al., 2010) lists 110 parameters which could be recorded. In the end, PROLOGUE will come up with a proposal about which of these parameters to include in a large-scale European ND experiment.

5 Previous Naturalistic Driving Research

PROLOGUE includes a comprehensive literature study about ND studies that have already been carried out so far (Backer-Grøndahl, 2009).

The "100 car study" (Dingus et al. 2006) was the pilot for a large-scale ND experiment in the US, which is currently being carried out within the Strategic Highway Research Program 2 (SHRP2). One hundred drivers who commuted on a regular basis in the Northern Virginia/Washington D.C. metropolitan area were recruited to the study. Seventy-eight drivers drove their own car, whereas 22 drivers used leased cars. All cars were instrumented, and data was collected over an 18-month period. Drivers using their own car received $125 per month and a bonus at the end, whereas drivers using a leased car received free use of the car and a bonus at the end.

All vehicles were instrumented with a package engineered by Virginia Tech Transportation Institution (VTTI). The instrumentation included the following: a vehicle network box that interacted with the vehicle network, an accelerometer box obtaining longitudinal and lateral kinematic information, a headway detection system providing information on leading or following vehicles, a side obstacle detection to detect lateral conflicts, an incident box, a video-based lane-tracking system and a video to validate any sensor based findings. The video-equipment consisted of five cameras monitoring the driver's face and driver side of the vehicle, the forward view, the rear view, the passenger side, and a view of the driver's hands and surrounding areas.

The data set included approximately 2,000,000 vehicle miles and 43,000 hours of data gathered over a period of 12-13 months for each vehicle. This research effort was initiated to provide an unprecedented level of detail concerning driver performance, behavior, environment, driving context and other factors that were associated with critical incidents, near-crashes and crashes. A primary goal was to provide vital exposure and pre-crash data necessary for understanding causes of crashes, supporting the development and refinement of crash avoidance countermeasures, and estimating the potential of these countermeasures in reducing crashes and their consequences.

A large-scale naturalistic driving study will be included as a part of the US Strategic Highway Research Program 2 (SHRP2). The in-vehicle driving behavior study will be conducted with volunteers driving instrumented vehicles for everyday use. An instrumentation package will be developed for installation on many vehicle models. The drivers will use their own vehicles during the study period. The driver and vehicle pool will change at least once a year through the reinstallation of the instrumentation package in a new driver's vehicle. The data collection package will accommodate requirements for a variety of analyses of lane departure, intersection crashes, and other questions.

The study will be conducted in several geographic areas to accommodate variations in weather, geographical features, and rural, suburban, and urban land use. Data will be archived for analysis as part of the data processing and will be made available to qualified researchers. Another critical need in the in-vehicle study of driving behavior is for detailed roadway data, with greater coverage of the roads used by the volunteer drivers. This data will support the association of driver behavior with roadway characteristics such as grade, curvature, and posted speed limits.

The field studies envisioned under SHRP2 will produce large data sets. Although data collection technology has advanced rapidly over the past few years, analytical methods have not kept pace. The field data collection projects are therefore supported by a series of projects at developing analytical methods. Key aspects of the analyses include the application of crash surrogate approaches, such as traffic conflicts, critical incidents, near-collisions, and other surrogate measures; development of exposure-based collision risk measures; and the formulation of analytical methods to quantify the relationship of human factors, driver behavior, and vehicle, roadway, and environmental factors to collision risk (Campbell and Mason, 2008). The main objective of this study was to investigate driver distraction among commercial vehicle operations using naturalistic driving data (Olson, Hanowski et al. 2009).

For the "heavy vehicle study", data from two previous naturalistic studies was combined; the two data sets represented 203 commercial vehicle drivers, and approximately 5 million km of data including kinematics and video data. The dynamic performance measures included longitudinal and lateral acceleration and braking, and the video included a view of the driver's face, as well as forward, backward left, and backward right road views. Trained data analysts identified so-called safety critical incidents defined as crash, near-crash, and crash-relevant conflicts. This was achieved by running the data through a software program in which incidents were flagged up based on certain thresholds: longitudinal acceleration (hard braking), time-to-collision, swerve, critical incident box, and analyst identified events. For each safety-critical incident, the analyst defined the type of conflict, potential distractions, driver behavior, and road and environmental conditions. 4452 safety-critical events were identified in the dataset, including 21 crashes. In order to estimate odds ratios, baseline events were also gathered, i.e., uneventful, routine driving (Olson, Hanowski et al. 2009).

Several other studies have been carried out using a ND approach or at least an approach which is similar to ND. The following issues have been predominantly researched:

- Driver distraction and inattention
- Drowsiness and fatigue

- In-vehicle systems
- Lane-change behavior
- Heavy vehicle – light vehicle interaction
- Driver characteristics and states

In addition to the fields listed above that all concern road safety, the following fields have to some degree been studied through use of naturalistic driving observation:

- Eco-driving
- Traffic flow/traffic management (lane change behavior)
- Relation between self-report and naturalistic observation

6 Methodological, Organizational and Ethical Issues

Besides the simply technical issues (also being part of the methodological ones), there are a number of questions to answer before an ND experiment can be started. First of all, the sample has to be dealt with. Sample size and representative nature are the key issues. Although naturalistic driving studies are typically observational, it still makes good sense to consider them in terms of experimental designs where an 'experimental group' is compared with a 'control group' in order to uncover the reasons why events did or did not happen.

Usually, there are large amounts of data to be processed in ND studies. A major issue here is to spot important events in the data so as to relate these events to interesting explanatory variables. It is desirable to develop an automated procedure to spot these events. There are statistical tools that can help with developing an efficient procedure that minimizes false positive events and also minimizes missed events.

Legal and ethical issues need close consideration. In particular, privacy and data protection are most likely to create problems, if they are not taken into account in the preparative phase of an experiment. It is reasonable that the driver receives a professionally prepared contract determining his rights and obligations. However, passengers in the vehicle have to be considered as well. Rules for recruitment and selection of test subjects have to be set up with regards to these issues and the necessities of scientific quality (Groenewoud et al., 2010).

Besides the logging and storing of data, there is another big technological issue and that is data processing. The huge amounts of data created by a ND experiment require appropriately extensive resources in terms of storage and processing, including both software and hardware. Automatic image recognition and processing including the triggering of particular events have to be taken into account.

In order to facilitate international cooperation, database management including exchange and transfer of data and interoperability as well as the flexibility to include additional parameters must be closely considered.

7 What Do Stakeholders Expect from ND Studies?

In order to address this question, PROLOGUE carried out a survey among the user group that was recruited during the first months of the project. The user group consists of representatives from governmental organization (European, national, regional and local government), the industry (automotive, supplier, insurance), from research organizations and from other organizations, such as police, road users, environment protection and driver training and testing organizations. A total of 137, mainly European, professionals in the area of road transport and related areas were invited by e-mail to participate in the survey. 72 people completed the questionnaire.

Almost all respondents held the opinion that ND studies would provide relevant knowledge about road safety issues. According to around one third of the respondents, it could also help to improve insight into human-machine interface design. Its usefulness for traffic management and particularly for environmental issues was less clear to the respondents. This assessment coincides with the way that ND studies have been applied so far. As concluded by Backer-Grøndahl et al. (2010), the vast majority of the previous and current ND studies have addressed road safety issues; only a few have addressed environmental or traffic management aspects. Since ND studies seem to be a very suitable method for collecting relevant information for these topics as well, it should be one of the additional objectives of PROLOGUE's dissemination work to clarify this more explicitly to future user groups.

From the current study it can be concluded that, overall, many topics are 'important' or 'very important' to investigate in a large-scale European ND study. According to the respondents, the four most popular topics are:

- Risk-taking behavior, in particular speeding, close following and drink driving
- Pre-crash behavior
- Crash avoidance behavior
- Driver condition, in particular fatigue and various health conditions

Fig. 1 shows the full set of ratings about the important topics of an ND study.

Implicitly, this implies which gaps in current road safety research can be filled by ND: the huge opportunity offered by ND is to take a big step beyond research on the basis of qualitative of quantitative accident data, i.e. from the collision itself and its outcome to the pre-crash phase, in addition to near misses, conflicts and normal driving behavior. It may help to determine where in this chain of events the path of an accident departs from normal driving, which may instead result in just a conflict or a near miss.

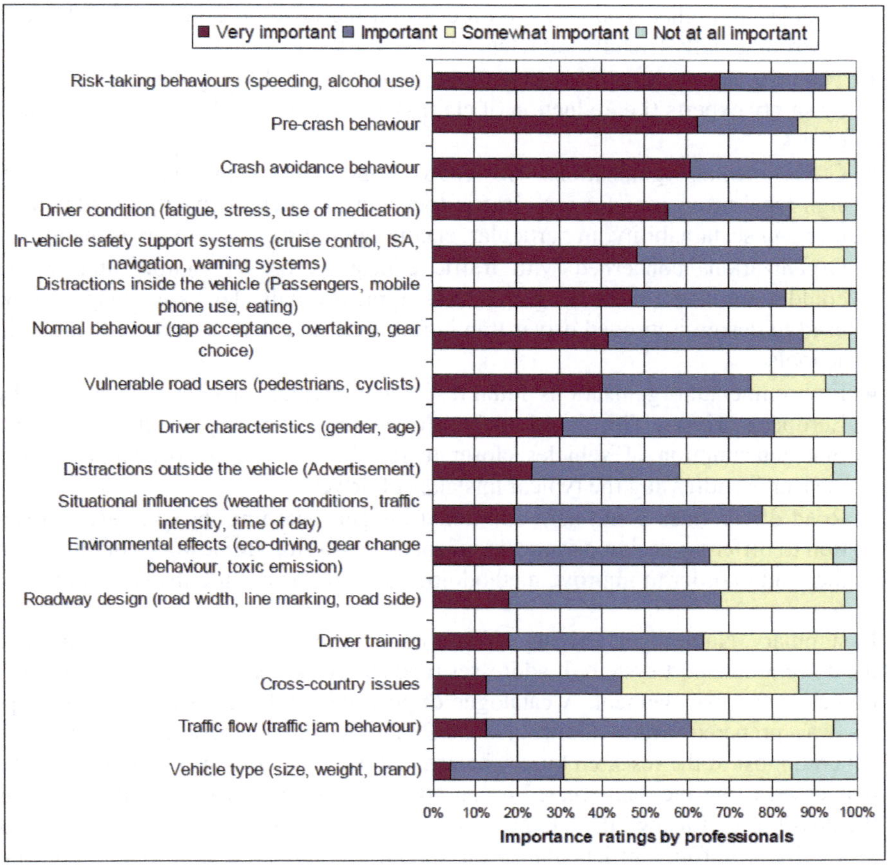

Fig. 1 Topics for an ND study ranked in terms of importance according to the respondents. (Source: Van Schagen et al., 2010)

8 Who May Benefit?

It is quite obvious that in the end, the road users shall benefit from a large scale European ND experiment. Data collection will establish the basis for a huge variety of research work later on. This may improve safe, comfortable and environmentally friendly mobility in the future. On the way there, there are a large range of stakeholders who may benefit from ND research. To give some examples:

- Road safety organizations may receive the tools to investigate driver behavior on an advanced level to gain insight into accident causation and be able to propose a next level of road safety measures.
- The car industry and its suppliers may gain additional knowledge of how their vehicles and equipment are used and experienced by both the users of their

vehicles and other road users in order to further develop safe, comfortable and environmentally friendly vehicles.
- The insurance industry may benefit from the new road safety measure proposed by safety experts (i.e. reduction of claims) as well as by enhanced predictability of risk.
- Governmental organizations will receive improved information in order to design legal provisions for road traffic focusing on safety, capacity of infrastructure and sustainability, in particular, encouraging infrastructure investments.
- Organizations concerned with traffic education, driver training and testing could learn where the average road user typically fails. This knowledge can be used to design improved driver training, or even to enhance traffic education in schools.
- Environmental organizations could research the typical driving style adopted by European drivers. This could enable the creation of new procedures to assess fuel consumption of vehicles closer to real-life driving or to design training measures addressing the typical mistakes of drivers.
- Road directorates could gain additional insight regarding the road users' reaction to different design parameters of roads or varying framework conditions on the road in order to improve methodologies for infrastructure management.

In summary, Naturalistic Driving offers a huge variety of opportunities to improve the current transport system. It addresses road safety as well as environment protection and road management. A catalogue of potential applications and research topics was prepared within the course of PROLOGUE (Sagberg et al., 2010). In general, most of the research topics, in terms of accident analysis in the past, can be reassessed using the data created by an ND experiment. But beyond that, there is the possibility of using ND data for researching the history of a collision as well as near misses, traffic conflicts, dangerous situations and "normal" driving behavior. More knowledge about normal driving behavior may provide additional input when designing fuel-efficient vehicles, which take account of this normal behavior. Furthermore, in traffic management normal driving behavior is a big issue, but also driving behavior within exceptional situation like traffic jams may be of interest. ND offers an opportunity to collect some exposure data and safety performance indicators. The EC-funded project DaCoTa (Road Safety Data Collection, Transfer and Analysis, some information is available at http://www.dacota-project.eu) has a work package on ND, where such issues will be dealt with.

ND as new methodology is also of particular interest for research on powered two-wheeler (PTW) safety. Rider behavior is hardly studied at the moment and there are a large number of unsolved problem and questions in terms of PTW safety, where ND is supposed to a reasonable approach to find solutions and answers. Both instrumented motorcycles and instrumented cars can be used to research, e.g. the problem of perception of PTW by other road users. 2 BE SAFE (Powered two wheeler behavior and safety, see www.2besafe.eu), another research project funded by the European Commission, will execute four pilot studies on naturalistic riding using instrumented motorcycles as well as an instrumented car.

A sophisticated ND experiment is too much of an effort to be carried out on a small scale. A European ND experiment would create comparable data throughout the European Union, being representative for the whole of the European population.

9 Next Steps

The most demanding part of the PROLOGUE project is to execute five pilot studies. These small-scale field trials shall:

- demonstrate the potential usefulness of naturalistic observations for various aspects of road safety through fundamental and applied research
- serve as pilots for a future large-scale naturalistic study by revealing the strengths and weaknesses of the data collected by the various instruments used, by identifying solutions for potential difficulties with data collection and data analysis.

The field trials are currently being prepared (Lotan et al., 2010). They will be carried out in five different areas:

Israel

This trial is aimed at extracting and understanding driving profiles of drivers through the use of data from two types of in-vehicle-data-recorders (G-based and vision based systems) with emphasis on young drivers. The Israeli field trial will use the "Green Box" technology. The data logger used is able to detect up to 110 different typical driving events. This has so far been used for training purposes, in particular of young drivers. The drivers get particular feedback, displayed as green, yellow and red boxes in a diagram.

The device used for this purposes can also be used to collect data in the sense of naturalistic driving. The algorithms developed for this application are a good starting point for developing algorithms for data mining and assessment for a large scale ND study.

Austria

This field trial seeks to monitor novice driver behavior from the learner period up to three months after passing the driving test with an in-vehicle device, which captures audio and video signals as well as acceleration and deceleration forces.

The "p-Drive" main unit contains all the acceleration and gyro sensors as well as a camera. It is normally used for driver training and retraining activities. In the case of PROLOGUE, it will not be mounted on the wind shield, which would interfere with the goal of unobtrusive observation. Instead, two "finger" cameras will be used and carefully hidden behind the wind shield. One camera will be directed to the front of the vehicle, the other to the driver. A GPS module will be linked. Decisions about additional sensors and CAN to be linked to the data logger are not yet settled. The videos will be recorded event-based (not full-time) in order to save disc space and extend the observational period. A combination of parameters to define such an event will be developed (trigger).

The Netherlands

This trial includes a site-based observation as well as in-vehicle observations. The site-based observation with video cameras allows the observation of all traffic passing at a given road location, including vulnerable road users such as cyclists and pedestrians. The purpose is to learn more about the interaction between different road users at intersections. Another objective is to explore the possibilities and opportunities of combining a site-based observation with an in-vehicle observation.

Spain

This trial is aimed at studying the use of nomadic devices while driving as well as evaluating the potential risk associated with the use of these devices. Large quantities of information are collected from the driver as well as from the vehicle and the road and traffic conditions using the ARGOS instrumented car. The ARGOS vehicle records images with nine cameras as well as sounds of the surrounding environment and the driver.

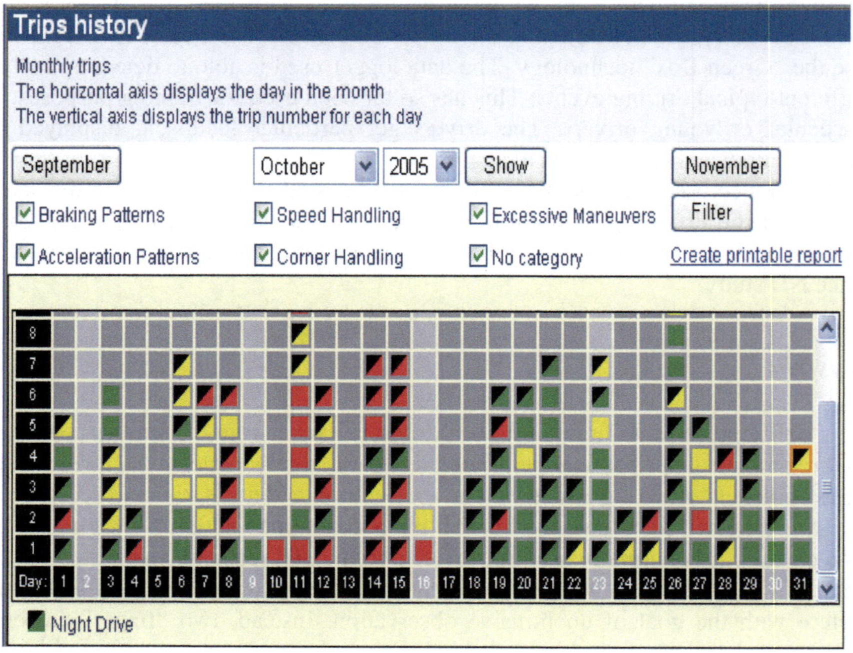

Fig. 2 Driver Feedback diagram for GreenBox by Greenroad Technologies

Fig. 3 p-drive main unit

Greece

This trial aims to identify changes in driving behavior when using a Forward Collision Warning (FCW) system and a Lane Departure Warning (LDW) system.

From the experience of the field trials and the theoretical research work carried out in the previous work packages, PROLOGUE will develop recommendations for how to carry out a large-scale, European Naturalistic Driving study.

References

1. Backer-Grøndahl, A., Phillips, R., Sagberg, F., Touliou, K., Gatscha, M.: Naturalistic driving observation: Topics and applications of previous and current naturalistic studies. PROLOGUE Deliverable D1.1. TØI Institute of Transport Economics, Oslo, Norway (2009)
2. Campbell, K., Mason, L.: Developing measures to improve highway safety. The safety focus area of the strategic highway research program 2. TR News (2008)
3. Dingus, T.A., Klauer, S.G., Neale, V.L., Petersen, A., Lee, S.E., Sudweeks, et al.: The 100-car naturalistic driving study, phase II – results of the 100-car field experiment. Report no. DOT HS 810 593, National Highway Traffic Safety Administration (NHTSA), Washington, D.C. (2006)

4. Groenewoud, C., Schoen, E., Malone, K., Jonkers, E., Hoedemaeker, M., Hogema, J.: Methodological and organizational issues and requirements for ND studies. PROLOGUE Deliverable D2.2. TNO, Soesterberg, The Netherlands (2010)
5. Llaneras, R.E., Freedman, M., Steinberg, G.V., Perel, M.: Naturalistic driving studies. A toll for the development and evaluation of in-vehicle systems (1999)
6. Lotan, T., Grimberg, E., et al.: Specification and Planning of Field Trials. PRO-LOGUE Deliverable D3.1. Or Yarok, Hod Hasharon, Israel (2010)
7. Olson, R.L., Hanowski, R.J., Hickman, J.S., Bocanegra, J.: Driver distraction in commercial vehicle operations. Report No: FMCSA-RRR-09-042. U.S. Department of Transportation, Federal Motor Carrier Safety Administration, Washington D.C. (2009)
8. Sagberg, F., Backer-Grøndahl, A.: A catalogue of applications and research topics for future naturalistic driving studies. PROLOGUE Deliverable D1.3. Oslo, Norway: TØI Institute of Transport Economics (2010)
9. Van Schagen, I., De Craen, S., Van Nes, N., Eenink, R.: The areas of interest of potential users for naturalistic observations studies. PROLOGUE Deliverable D1.2. SWOV Institute for Road Safety Research, Leidschendam, The Netherlands (2010)
10. Welsh, R., Reed, S., Talbot, R., Morris, A.: Data collection, analysis methods and equipment for naturalistic studies and requirements for the different application areas. EC FP7 project PROLOGUE Deliverable D2.1 (2010)

Re-thinking Urban Mobility Services and Operations

An Enhanced Individual and User Specific Service Concept for Urban Mobility

Kyandoghere Kyamakya and Oana Mitrea

Abstract. Recent data on urban traffic mobility emphasizes the difficulty of the current planning strategies in reaching the goals of either reducing, shifting or avoiding heavy traffic that is caused primarily by individual private vehicles. An interdisciplinary approach to designing a highly efficient, intelligent traffic and transportation concept is required, in which the development of innovative services and operation solutions is combined with a closer examination of the various user and stakeholders needs, their strategies as well as their dynamic grade-of-service requirements to cope with daily movements and accessibility. Considerations from the field of science and technology social studies will be combined with the critical analysis of the current urban mobility solutions. This knowledge is then used in a design process to create a sustainable example of an urban mobility service concept with the focus on both inter-modal and inter-service convergence, which we consider to be particularly attuned to the preferences and handling modes of the target users: a "Mobility stock market".

Kyandoghere Kyamakya
Transportation Informatics Group,
Institute of Smart Systems Technologies,
Alpen-Adria University Klagenfurt, Klagenfurt, Austria
e-mail: `kyandoghere.kyamakya@uni-klu.ac.at`

Oana Mitrea
IFZ - Inter-University Research Centre for Technology,
Work and Culture Graz/Klagenfurt,
Department of Research on Science and Technology,
University of Klagenfurt. Sterneckstrasse 15, 9010, Klagenfurt, Austria
e-mail: `oana.mitrea@uni-klu.ac.at`

J. Düh et al. (Eds.): Data and Mobility, AISC 81, pp. 177–186, 2010.
springerlink.com © Springer-Verlag Berlin Heidelberg 2010

1 Intelligent Transportation Systems as Socio-technical Heterogeneous Constellations

In recent years, the need for mobile access to communications and to data services - anytime, anywhere - has become increasingly clear and 'smart communication' has seen rapid acceptance and growth. Intelligent internet-based communication and mobility devices and systems are rapidly improving and changing, resulting in a growing demand for a new kind of data access: users are now looking for means to access their information while on-the-move. At the same time, they want to use this to dynamically and efficiently coordinate their increasingly mobile and fragmented daily life.

The new intelligent transportation systems have grown into a nerve centre of urban activities integrating movement, communication and information into a complex structure. How and how much do people communicate while moving, and move while communicating, and how do the telecommunications/movement practices change the way people live, interact, and work? These are issues that closely intertwine with the requirements and functionality of the current intelligent mobility systems.

What does *intelligent* mean in relation to transportation? Starting from the observation that "ITS is a dynamic field, (...) and many of its research streams involve pervasive computing research and applications" (Wang et al. 2006, p. 68 [1]), their "smartness" is mainly related to ambient intelligence solutions such as:

- *the active in- and -out vehicle environment sensing*, or road infrastructure sensing which will provide users with improved safety, spontaneity, efficiency and comfort (Wang et al. 2006, p. 68 [1]);
- *intelligent spaces defined as "environments that continuously monitor what's happening in them, communicate with their inhabitants and neighborhoods, make related decisions, and act on these decisions" (Wang et al. 2006, p. 68 [1])*
- *agent-based control – distributed operational agents letting networked transportation systems operate on a management-on-demand or service-on-demand basis (Wang et al. 2006, p. 69).*

During usage, humans combine existing applications and arrangements in a sometimes unexpected, creative way - which we can also call "smart". Their actions are not linear or easy to understand and predict, but are trade-offs dependent on a variety of social and technical variables, which work together in "constellations" or "configurations". Such a heterogeneous constellation is, for instance, the *intelligent traveling space* responsible for monitoring, communication and decision-making capabilities which interact with the mobility behavior of travelers by observing them and checking this according to specific rules inscribed in the system. The term '*heterogeneous constellations*' is introduced here in accordance with the latest findings from studies of science and technology analyzing the agency of humans and intelligent technologies in hybrid systems (Rammert 2007 [3]). They raise the question as to what extent well-known interaction concepts specific to social theories can also be applied to the analysis of the interaction in these systems (Weyer 2009, 75 [2]).

As Schultz-Schaeffer argues, we are used to seeing the relation between users and technology as one of command and execution – technical device/systems being mainly regarded as operative, passive means of carrying out actions. In reality, the relation between users and technology is much more complex, also concerning the simplest technologies (Schultz-Schaeffer 2009, 39 [4]), because technical devices and systems can *behave* in such a way that they enable only specific types of actions of users (*id.* [4]). The cited author further highlights that the software modeling of decision and coordination processes, and the development of interactive interfaces between technologies and their users lead to a further blurring of the borderline between the "disposed" user and the "executing" technology (Schultz-Schaeffer 2009, 40 [4]). The new developments in robotics and agent technology endow such advanced technologies with a greater flexibility of self-controlled behavior, a higher capacity to interact with human users and other technological components. In this way, technology does not represent just an isolated, fixed and efficient means of action in an agency chain exclusively controlled by users, but a mesh of activities and interactions between humans, things and symbols (Schultz-Schaeffer 2009, 40 [4]).

To understand heterogeneous constellations involving humans as users and developers and other interacting technologies, we aim at looking at how agency is distributed between technical and human actors (Schultz-Schäffer 2009 [4]). This means that at least some parts of actions and/or *arrangements* are taken over by the agency capacity of ITSs (*id.*, p. 50 [4]). They mainly refer to *connection, real-time reaction, and situational behavior*. Particular attention should be paid in this context to the *possible embedding of processes of categorization, selection and exclusion* into the system.

Recent sociological studies of technology emphasize the fact that the distribution of activities among humans and machines in hybrid socio-technical constellations has important consequences for system design (Rammert 2007, 130 [3]). This implies that several design criteria such as: affordability, controllability, safety, pleasure of driving and privacy should be carefully interpreted and implemented, so that a balance between the distributed human and technological actions can be reached (*id.* [3]) The self-initiative, responsibility, control of personal data, intervention capacity and the decision about the real usefulness of applications (Rammert 2007, 131 [3]) should further characterize human actions within intelligent transportation systems.

All in all, the development of intelligent transportation systems in general should simultaneously consider the *rising 'agency' of technological artifacts and* the *distribution of activities in hybrid constellations*. The co-agency capacities of the technological frames and elements (in terms of real-time reaction, situational behavior, inter-modal and inter-service selection, scheduling, routing, accounting, and payment, and profile management) should be placed in relation to variables defining mobility and communication actions of users (socio-demographic variables defining various user groups, types of communication/information behaviors, types of social interaction, places to use communication technology for mobility purposes and mobility behavior variables such as travel purposes, frequency of voyages, travel distances, travel time, perceptions of own mobility, orientation skills, and effects of wireless technology usage on movement).

2 Problems of Developing User Specific Mobility Services in the Urban Area

We employ the term urban mobility in the sense of city transport/traffic movement designating "practices, conditions and services that allow movement from one place to another within the urban area" (Project SKILLS 2006, 27 [5]). This can include both human movement and city logistics related mobility. Although one may think that city logistics and human traffic movement represent separate fields, recently logistics services are also becoming increasingly based on the building of partnerships and the introduction of new forms of cooperation between all actors those involved with the delivery /picking up of goods within cities. These partnerships offer a significant reduction in vehicle miles and number of trucks, thus reducing traffic congestion.

The impact of specific user services for mobility on urban development should be taken into consideration in any development model of sustainable European cities. Recent studies about intelligent transportation systems stress the necessity of formulating policies addressing both system and economic-social inefficiencies (Kanninen 1996 [6]). According to a policy formulation by the European Commission from 2003, for instance, Intelligent Transportation Systems should accomplish the following objectives: offer improved information about public transportation: routes, plans, prices, enabling the route planning in the city (including parking), supporting alternative means through real-time information about cycling lanes, offering recommendations for route changes during travel by means of real-time information on mobile phones or screens, providing payment modalities for traffic services on mobile phone or by means of smart-cards (European Commission, 2003 [7]).

Within the sustainability paradigm, the effects of the latest traffic solutions based on information and communication technologies cannot be punctual because of their complexity and rapid interaction capability with other infrastructures and systems. In connection with city sustainability, it is mainly their systemic effects which are addressed, in the light of their potential as means of responsible development across the entire urban area. More than just simply fulfilling the mobility needs of specific user categories (particularly from disadvantaged areas), the hopes are that efficient and affordable transportation systems would also improve social cohesion, communication and information, increase self-organization, diversity in society, economy and culture, enhance trust and cooperation through networks and encourage participation - all dimensions of social capital as 'one of the central resources of sustainable development' (Schwarzer et al. 2006, p. 12[8], see also Petrakis et al., Eds., 2007 [9]).

It should be admitted that there are still problems with the accomplishment of basic user mobility needs in the different city areas. Users (persons or organization that buys transportation services) and their needs are fundamental to the existence of transportation organizations. Thinking about what is important to the user/customer, from the perspective of what services transportation organization provide is fundamental. The choice of how to travel will often vary as a function of the purpose of the trip. One should distinguish between privately owned vehicles such as automobiles and publicly owned vehicles such as buses and trains. Some Grade-of-Service variables for travelers are: *price or fare, travel time, service frequency, comfort, service flexibility*, etc. Customers (either shipping freight or people) have a choice among different modes and they have a choice about carriers within a particular mode. The choice a customer makes will presumably be in their own self-interest: *they will optimize a set of variables by studying the grade-of-service for various options.*

In transportation, the grade-of-service is thus multidimensional. It represents a combination of a number of the different variables, indicated above, that people (i.e. travelers, shippers, receivers) will integrate and internalize in order to make a judgment about what mode and carrier to select. For the purpose of analysis, one often needs to reduce the multi-dimensional grade-of-service it to a one-dimensional variable. Thereby one needs a way to collapse these variables into a single one. So we define a measure called "UTILS" – for "UTILITY" which is a linear and weighted combination of the different grade-of-service variables. The weighting coefficients should reflect the relative importance of each of the variable to the customer. Also the sign of these weighting coefficients is also important; it is negative for those grade-of-service variable for which "the more, the worst" is true. For example, the variable "travel time" should have a negative weighting coefficient.

Suppose we have three possible modes or service providers within a mode. We can compute the UTILITY for each of the service options and will then select the one offering the highest UTILILITY. The UTILITY is subjective to a customer since the weighting coefficients reflects what he/she personally does expects in term of grade-of-service.

Because recent market developments have brought a wide range of intelligent mobility solutions to the urban area: from the classical solutions applying to individual transportation and public transportation (bus, tram, taxi) to the newer solutions - allowing car sharing, hailed shared taxis, bus-on-demand, car renting, etc., we should first formulate the core requirements of GOS expressed by each of the players in order to judge the different mobility options currently offered in the urban context.

Consider the passenger/traveler: their views of the grade-of-service (GOS) are: time flexibility, space access or destination flexibility, shortest travel times, lowest cost, sufficient comfort, information about the service quality or dynamics during both the pre-trip and on-trip phases. This grade of service reflects many inherent problems that are: (a) it is multi-dimensional, therefore it will not always be possible to ensure the highest level for each of the dimensions at the same time (multi-dimensionality of the GOS); (b) not all users will have the same GOS requirement levels at any time and at any location (diversity of GOS levels amongst travelers/customers).

Consider the transportation services operators/providers: their view of the grade-of-service, i.e. from their point of view, is to provide, globally (i.e. the average/median), the highest possible grade-of-service to the customers/travelers at the lowest possible operational cost and therefore ensure the highest possible profit - if any at all. The profit related aspect cannot be neglected since it has to do with potential incentives for novel business ideas that will help solve the current challenging problems in this sector.

Consider society in general: as Weyer emphasizes, in the construction and implementation of telematic traffic control systems, questions arise which are discussed in sociology under the label: *market versus state*. This points towards conflicts between individual and collective interests; and local and global optimization (Weyer 2009, 70 [2]). In the light of these considerations, the commonly held perspective is that the highest possible grade-of-service to the customers/travelers should be guaranteed, while keeping the overall costs of the mobility as low as possible, mainly as regards negative impact on the environment. The latest ideas include aspects such as emissions and air pollution, noise, congestion, etc. This requires the harmonization of individual and collective; global and local interests.

Having assessed the core GOS requirements of the different players, we can now briefly analyze to what extent the different current mobility solutions face serious problems in satisfying these challenging requirements in a trade-off-manner while taking all stake-holders into account globally. Let us select a representative sample of the most prominent solutions: (1) taxi; (2) privately owned car; (4) car rental; (4) car-sharing; (5) paratransit services; and (5) public buses and trams.

Most of these listed services fail to realize the trade-off concerning a global comprise of the GOS as respectively viewed by each of the stake-holders. The taxi does offer a high GOS for most dimensions except for the fare. The fare is high and there is only one service class offered, a premium one. Conventional fixed route buses and rail modes are too expensive, inefficient or inflexible in the low traffic time windows such as weekends, night times and on holidays. They are the best in terms of low fares but offer a relatively poor GOS in the other dimensions. The fare here is the lowest and there is only one service class offered, i.e. a low-class one. The paratransit and the car-sharing concepts appear to come much closer to a good global trade-off. However, they are not capable of maintaining an optimal trade-off while facing a fluctuating dynamic demand, as is the case in reality. In summary, not one of the approaches is perfect despite the fact that some of the concepts seem closer to a trade-off. It is also very important to notice that the GOS level from the point of view of a given stake-holder may depend on the actual demand/load. For example, from the operator's point of view, a fixed line bus service displays a very low GOS in low-traffic time-windows like nighttime, weekends and on holidays. On the other hand, for example, a taxi system may offer a lower GOS to the travelers in case of extremely high demand: customers may wait too long – in some cases up to hours before getting a taxi.

Since a single mobility solution alone cannot keep the GOS consistently high for all stake-holders under all traffic/demand load levels, a combination and/or convergence of different solutions will be the optimal concept. This virtual integration of diverse mobility solutions should include a GOS - aware complementarity in both time and space. Furthermore, to save costs, different GOS-level dependent service classes should be defined and pay different fairs accordingly; the premium class(es) will enjoy more priority but pay more than the others.

Optimized and efficient access to information is in fact the basis for any service supporting the urban mobility related processes. All observations support the idea that mobility services featuring optimized multimodal and multi-service integration have great potential for all stakeholders with regard to environmental sustainability and business constraints (Litman 2009 [10]). Intelligent mobility services are relatively new and, although proven technically, they still present challenges, especially in terms of matching strategic objectives with assured delivery. To enable multimodal and integrated intelligent mobility services with clear benefits for the customers, the city, the mobility related stake-holders, and the wider economy, the urban mobility services related ITS projects and plans need to be integrated at various levels to show how individual projects contribute to the delivery of a broad ITS strategy (see the broad set of dimensions listed in the related maturity model presented by Houghton et al (Houghton et al. 2009 [11]). The strategies for the city's different mobility services categories and modes should be integrated in a coherent model, which is also consistent with the measures taken in other areas of municipal government (for example, land-use planning).

3 The Service and Operation Concept Mo-Bay (Mobility Stock-Market)

The service and operation concept developed by The Transportation Informatics Group at the University of Klagenfurt (Prof. Dr.-Ing. Kyamakya) has the pretention to be similar to something like a middleware that integrates virtually all mobility services that are on the market. Its innovation is expressed by the following unique features: (1) involvement of private and rental cars that may offer to transport some people during their normal journeys– this will be enabled by an intelligent mobility service with respect to data/information visibility, and dispatching; (2) dynamic, that is demand versus resources -aware responsive pricing for different GOS classes; (3) possibility to have a smooth intermodal and inter-service routing; (4) it bases scheduling, and routing decisions on real actual data that is collected or provisioned via the mobile service; (5) automatic accounting and cash-free payment.

The concept uses the intelligent information system technologies to coordinate mobility needs, mobility demand, goals and actions in real-time through the efficient combination of multiple urban modes of transportation (individual cars offering lifts on their normal journeys and public transportation methods (i.e. bus, metro, trams, and taxis). This enables both the integration and also the dynamic optimization of a virtual mobility that will use and integrate as much as possible the available mobility systems, infrastructure and capacities. This means that the highest possible grade-of-service at the lowest possible cost will be the permanent target of all real-time optimization processes. The pricing will be also dynamic, reflecting the actual ratio between available utility-grade-of-service (the offer and/the resources) on one side and the customized demand on the other.

A comprehensive consistent systems engineering process is followed. The systematic capture of all requirements while taking account of all stake holders will be one of the key challenges here. The following technical implementation is at least straightforward due to the maturity of the core technologies. The core of Mo-Bay will be an ICT platform that will enable a smooth integration of all interested service providers as well as the users and demanders. These items will, however need appropriate terminals and/or mobile devices.

After implementation of the technical prototype, a test-bed will be conducted with a selected number of players (service providers and users) in order to systematically test it according to the core of the requirements specified in the engineering documents. Beyond the technical view of the prototype system, grade-of-service related issues will be analyzed, and dynamic pricing concepts will be fine-tuned.

The development and piloting activities are accompanied and shortly followed by the intermediate analysis of the impact of the new mobility solutions on urban development. The objectives are to make comparative assessments of alternative projects and financial measures, to investigate possible synergetic effects of the new transportation model in the urban area, to analyze its interaction with other urban infrastructures/subsystems and to assess the risks/ undesired systemic effects. We employ a method making use of several tools: an impact map to clarify the effects to be evaluated and qualitative interviews with the involved stakeholders.

4 Conclusion

If we regard the **Mo-Bay** concept from the perspective of its structuring effects for both communication and mobility, its flexible, interactive arrangements for intermodal and inter-service selection, scheduling, routing, accounting, and payment seem particularly suitable for the fragmented, non-linear way that users of information and mobility in society behave: when somebody moves and simultaneously uses wireless communication/information devices on-the-move, place-related communication (appointments, meetings, meeting places, requirements to solve problems appearing “on-the-move”) serves to organize users’ mobility according to various mobility coordinates. In turn, mobility leads to a strong fragmentation of viewpoints and experiences in the exchanged content, meaning that users flexibly and dynamically react to various problems occurring in ever-changing contexts. They finally decide about transportation taking into consideration a variety of grade-of-service variables. It should be said that the grade-of-service itself is differently perceived and defined by the various players in the system, therefore making the harmonization of individual and collective; respectively global and local interests highly important.

Another argument regarding flexibility: the intelligent mobility service enables the involvement of private and rental cars that may offer to transport some people on the way to their normal journeys. This aspect challenges the concept of the passive user in the sense that travelers are seen as both providers and beneficiaries of the system, being particularly familiar with problems and advantages of both sides.

As Weyer emphasized, (in relation to navigation devices) sociology itself should develop new models to understand the new forms of socio-technical interaction (Weyer 2009, 70 [2]). Since the current mobility service and operation concept includes a great deal of suppositions about social behavior and social structures, we completely agree with the author's opinion that sociology could really make an important contribution to the understanding and development of such concepts (Weyer 2009 [2]).

References

1. Wang, F., Zeng, D., Yang, L.: Smart Cars on Smart Roads: An IEEE Intelligent Transportation Systems Society Update. IEEE Pervasive Computing, 68–69 (October-December 2006)
2. Weyer, J.: Die Kooperation menschlicher Akteure und nicht menschlicher Agenten. Ansatzpunkte einer Soziologie hybrider Systeme. In: Berger, W., Getzinger, G. (eds.) Das Tätigsein der Dinge. Beiträge zur Handlungsträgerschaft der Technik, Profil, München, Wien (2009)
3. Rammert, W.: Technik – Handeln – Wissen: zu einer pragmatischer Technik- und Sozialtheorie. VS Verlag für Sozialwissenschaften, Wiesbaden (2007)
4. Schulz-Schaeffer, I.: Die Frage nach der Handlungsträgerschaft von Technik und das Konzept graduelisierten Handelns. In: Berger, W., Getzinger, G. (eds.) Das Tätigsein der Dinge, Beiträge zur Handlungsträgerschaft der Technik, Profil, München, Wien (2009)
5. (2006) SKILLS. Final Report, The URBACT Programme, `http://urbact.eu/fileadmin/general_library/Building_Sustainable_Communities_-Final_Report_01.pdf` (accessed November 12, 2009)
6. Kanninen, B.J.: Intelligent transportation systems: An economic and environmental policy assessment. Transportation Research Part A: Policy and Practice 30(1), 1–10 (1996)
7. Europäische Kommission GD Energie und Verkehr: Intelligente Verkehrssysteme- Intelligente Technik im Dienst der Verkehrsnetze. Amt für amtliche Veröffentlichungen der Europäischen Gemeinschaften, Louxemburg (2003)
8. Schwarzer, S., Ornetzeder, M., Feichtinger, J.: Sozialkapital und Nachhaltige Entwicklung. In: WWW.Nachhaltigkeit.at. Internetportal zur Nachhaltigkeit in Österreich, Thema des Monats 6/2006, `http://www.nachhaltigkeit.at/reportagen.php3?id=5` (Accessed August 7, 2006)
9. Petrakis, K.N., et al. (eds.): The CITUM Guide for Urban Mobility Practice. The URBACT Programme (2007), `http://urbact.eu/fileadmin/general_library/CITUM_Guide_for_Urban_Mobility_Practices_20070729.pdf` (accessed November 12, 2007)

10. Litman, T.: Introduction to Multi-Modal Transportation Planning. Principles and Practices. Victoria Transport Policy Institute (2009), http://www.vtpi.org/multimodal_planning.pdf (accessed May 5, 2010)
11. Houghton, J., Reiners, J., Lim, C.: Intelligent transport. How cities can improve mobility. IBM Institute for Business Value (1999)

Road Weather Information Service in Finland

The Impact of Error Probability on Service Benefits

Risto Öörni, Satu Innamaa, and Risto Kulmala

Abstract. Road weather information services are provided to road users to improve their safety and mobility. In Finland, the information on current and forecasted road weather has been available to road users since the 1990s and along with developments in technology and service provision, the quality of the information provided to the road users has improved over time. While the overall successfulness of the road weather information service has been analyzed during several recent winter seasons, the impact of the improved service quality on the socio-economical benefits of the service has not been analyzed. The objective of this study was to analyze the relationship between the changes in service quality and the socio-economical benefits. The focus of the various elements of service quality was set on the veracity of the information, which is defined as the error probability in this case. The service benefits were quantified on the basis of the relationship between service output and behavioral reaction-related impacts on safety. In addition, the costs of the service were described at two levels of service quality.

1 Introduction

Road weather information services are provided to road users to improve their safety and mobility. In Finland, the information on current and forecasted road weather classification (normal/poor/hazardous) has been available to road users since the 1990s. Along with developments in technology and service provision, the quality of the information provided to the road users has improved over time. The overall success of the road weather information service has been analyzed during several recent winter seasons; however, the impact of this improved service quality on the socio-economical benefits of the service has not been analyzed.

Risto Öörni · Satu Innamaa · Risto Kulmala
VTT Technical Research Centre of Finland
P.O. Box 1000, FI-02044 VTT, Finland

J. Düh et al. (Eds.): Data and Mobility, AISC 81, pp. 187–201, 2010.
springerlink.com © Springer-Verlag Berlin Heidelberg 2010

Weather-related information and warning services are provided to road users in many countries. In most cases, the objective of these services is to improve the safety of road users by informing them about potentially hazardous conditions on the road network. The practical implementation of these services differs from one country to another which means that there are also differences in the quality of service experienced by the users. So far, there have been no commonly available tools to find the optimum service quality for these services.

The concepts of service quality and data quality have been studied in the European QUANTIS project (quality assessment and assurance methodology for traffic data and information services) (Anonymous 2010). The objective of the QUANTIS project is to study the concepts of data and service quality, develop a methodology to find the optimum service quality for different traffic information services and to apply the methodology in four European service cases. The use of road weather information in various applications such as traffic information services is one of the service cases undergoing analysis in QUANTIS.

The concept of service quality is defined in QUANTIS as a combination of six quality elements (timeliness, completeness, availability, veracity, precision and relevance) (Öörni et al 2009). Each of these six quality elements are related to one or more quality attributes which can be used as quality indicators for the output of an information service.

Finding the optimum service quality for any traffic information service is important for several reasons. First, providing any traffic information service usually requires considerable investments in service provision. Without knowing the service impacts and costs at different levels of service quality, one cannot find the quality level best suited for his or her objectives and make informed decisions related to service provision and development. This means that one may not be able to choose the deployment option leading to the optimum level of service quality. Second, limited availability of information may also be a barrier to investments in service development and service provision.

2 Objectives

The objective is to study the optimum level of service quality for the Finnish road weather information service and to test the methodology to find the optimum service quality developed in the QUANTIS project. The quality attribute under analysis is the error probability of the service output. Its relationship to service benefits and costs will be analyzed in this paper.

3 Methods

3.1 Finnish Road Weather Information Service

The Finnish road weather service is a traffic information service that provides road users with information on predicted road weather conditions via the internet and as part of weather forecasts broadcast on television and radio. The service collects and combines data on road weather, road maintenance and current weather, and forecasts the development of road and weather conditions based on this data for the next 24 hours. The forecast is given in three classes: normal, poor or hazardous. Road weather is typically classified as poor for 20–30% of the wintertime and hazardous for 5% of the wintertime. The forecast is updated four times a day.

Road weather information is predicted in the following way: first, road maintenance give their report on the current road weather and a proposal for road weather information for the next 6 hours. They base their report on the weather information and forecasts of Foreca ltd., the satellite and weather radar information of Finnish Meteorological Institute and on real time information from the fixed road weather measurement stations and fixed road weather cameras of the Finnish Transport Agency.

Second, the four traffic management centers of the Finnish Transport Agency compile road weather information at a provincial level for the next 6 hours. The road weather for the province is determined as the worst road weather on any of the main road sections within the province. Finally, the Finnish Meteorological Institute puts the information together and extends the forecast period to 24 hours. They may select a worse road weather classification than proposed by the Finnish Transport Agency if the weather forecasts necessitate this or if information for an area larger than a province differs greatly.

3.2 Service Quality

The concept of service quality can be defined as the quality of the information provided by an information service. The quality of information is analyzed at the point where it leaves the service platform operated by the service provider (Figure 1).

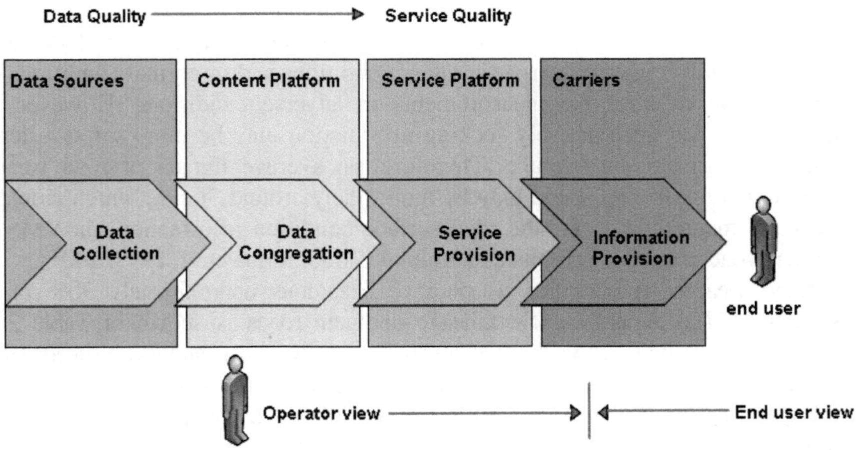

Fig. 1 Parts of the service production process (Scheider et al. 2010)

For the evaluation of optimal quality levels, the method developed in the QUANTIS project will focus on the service output. Due to the variety of communication media and end-user equipment, it is too extensive to assess the service quality at user presentation level. The output at service platform level is still very similar and is not restricted by the quality of presentation or communication technology, but rather is composed at an information level rather than a data level. This makes it possible to establish a more direct link between service quality and its socio-economic benefits. (Scheider et al. 2010)

Service quality can be defined as a combination of six quality elements: veracity, completeness, timeliness, availability, consistency and relevancy. In this case, the veracity of information provided to the end-user is measured in terms of error probability. The definition used for error probability is based on ISO technical report ISO/TR 21707 which defines error probability as a percentage of service output outside the stated quality boundaries (Anonymous 2008).

To find the optimum quality level for a service, a cost-benefit analysis should be performed for service at various levels of service quality.

4 Results

4.1 Socio-economic Impacts of Current Service

Socio-economic impacts of the weather information services were studied here from three different points of view: based on impacts on (1) travel time and its predictability, (2) safety as well as on (3) valuations and comfort (Table 1).

The average speed (km/h) was used as an indicator for travel time and predictability. The change in average speed has an impact on time costs. Sihvola and Rämä (2008) studied the impacts of road weather forecasts. According to their study when the road weather forecast was classified as "poor", those who had received road weather information drove 5km/h slower than others (speed limit 80km/h). The reduction in average speed differed between 5–10% when comparing traffic in adverse road weather conditions and in normal conditions one week before or after the measurements in adverse conditions. However, the drivers who had been actively seeking information may be more safety oriented than other drivers, on average. Therefore, an average impact of road weather information on driving speed may be more likely around 3km/h, which would be the same magnitude as for the road surface condition information via VMS for slippery road conditions (Rämä and Kulmala 2000, Rämä et al. 1999, 2001).

Road weather is classified as poor or hazardous approximately 30% of the wintertime. The annual vehicle mileage on main roads is shown in Table 2 for different vehicle types, as well as the hourly rate of time, costs and mean speeds (Road Administration 2009 and 2005, Kangas and Kärki 2009). Approximately 45% of mileage is driven in wintertime (Peltola et al. 2004). The interviews carried out by Sihvola and Rämä (2008) showed that 62% of drivers had sought road weather information. If we apply this proportion to the whole driver population, the 3km/h speed reduction due to the service use results in time costs of 19,800 €.

Table 1 Indicators of socio-economic impacts of the weather information services. The most important indicators are marked with black and the fairly important ones with grey.

QUANTIS - Indicators of socio-economic benefits and their valuation in cost-benefit analysis; Finnish service case (road weather information services via mass media for road users)							
			Impacts on driving costs. Transport economy-related impacts with specified shadow prices		Market-price impacts. Impacts in monetary terms that can be estimated / calculated		
Overall objective	Indicator	Unit	accident costs	time costs	on the economic status of private persons (e.g. changes in service prices)	on corporate economy	Not to be included in cost-benefit analysis
Time and predictability	Travel time (average and standard deviation)	Hour					
	Total door-to-door travel time	Hour					
	Public transport's deviations from timetables						
	Spot speed (average and standard deviation)	Km/h					
	Vehicle km travelled in congestion	Veh-km or ton-km					
	Stability of traffic flow (number of changes in speed)	Number					
	Perceived fluency of traffic flow						
	Success of Information services						
	Average traffic flow	Veh/h					
	Average traffic speeds	km/h					
	Length of queues	Metres					
Safety	Number of Traffic Accidents (per traffic unit)	Number of accidents (per veh-km)					
	Number of Traffic Accidents Injuries (per traffic unit)	Number of injuries (per veh-km)					
	Number of Traffic Accidents Fatalities (per traffic unit)	Number of fatalities (per veh-km)					
	Vehicle KM driven	km					
	Number of conflicts (near accidents)	Number					
	Feeling of safety						
Valuations and comfort	Willingness to pay (services)	Monetary					
	Number of users of a service	Users					
	Travel comfort experienced by users						

Table 2 Time cost calculation for extra travel time due to the 3 km/h reduction of travel speed in poor or hazardous road weather conditions (Road Administration 2009 and 2005, Kangas and Kärki 2009).

Vehicle type	Vehicle mileage (veh. km)	Average speed (km/h)	Hourly time cost (€)	Time cost due to the service (€)
Passenger car	29,895,000	90.5	16.03	-15,200
Van	2,525,000	90.5	16.77	-1,300
Truck	2,745,000	82.2	18.30	-1,900
Bus	395,000	85.7	101.18	-1,400

In addition to the effect on journey time, the service enables its users to have more predictable travel times. These impacts in addition to the actual average travel time impacts will be reflected in the valuation and comfort of travelers, with these impacts being reflected in their willingness to pay, therefore we have assumed that the improvements in predictability will fully compensate for the longer journey times.

Safety was measured by the number of traffic accident fatalities, injuries and traffic accidents, as well as vehicle kilometers driven and with their impact on accident costs (Table 1). Note that 45% of crashes involving injury and 47% of fatal crashes occur during the winter season, which reflects the vehicle kilometer distribution quite well (Peltola et al., 2004). Safety impacts are assumed to have occurred on main roads.

On the basis of the studies carried out by Aittoniemi (2007) and Hautala and Leviäkangas (2007), the likely reduction of crashes involving injury in poor and hazardous road weather conditions on the main road network could be approx. 10% resulting from the road weather information service. Rämä et al. (2003) estimated that real-time VMS information on slipperiness and other road weather related problems reduces the risk of accidents involving injury in adverse conditions by 8% on main roads and 5% on minor roads in Nordic conditions. Road weather is classified as poor or hazardous approximately 30% of wintertime. The annual number of traffic accident injuries on main roads is approximately 2,160 and accidents involving injury, approximately 1,490 (year 2008) and 45% of accidents involving injury occur during the winter season (Peltola et al. 2004). Consequently, a reduction of 10% of injury accidents in adverse road weather conditions would mean a reduction of 28 traffic accident injuries and 20 ($\approx$ 1,490 * 0.45 * 0.30 * 0.10) accidents involving injury per year. The average cost of an accident involving injury is 330,000 € (Road Administration 2005). Consequently, the benefit of the road weather information service is 6,600,000 € in reducing accidents involving injury.

Hautala and Leviäkangas (2007) estimate that approximately half of the safety impact of weather information is due to speed reduction. Nilsson (2004) showed that the speed reduction's safety effect for fatal accidents is equal to the squared safety effect for crashes involving injury. On the basis of these studies, we can estimate the effects on fatal crashes in poor and hazardous conditions to be 15%. The annual number of traffic accident fatalities is approximately 120 and fatal road traffic accidents 110 (year 2008) and 47% of fatal crashes occur during the winter season (Peltola et al., 2004). Consequently, a reduction of 15% of fatal accidents in adverse road weather conditions would mean a reduction of 2.5 traffic accident fatalities and 2.3 ($\approx 110 * 0.47 * 0.30 * 0.15$) fatal road traffic accidents per year. The average cost of a fatal road traffic accident is 2,205,000 € (Road Administration 2005). Consequently, the benefit of the road weather information service is 5,071,500 € in the reduction of fatal accidents.

Hautala and Leviäkangas (2007) assessed 1–2% annual crash reduction as resulting from the road weather information service. The total number of crashes with no injuries was approximately 6,670 per year on main roads in 2008. Consequently, the annual reduction is approximately 100 road traffic accidents with no injuries. The average cost of such an accident is 2,700 € (Road Administration 2005). Consequently, the benefit of the road weather information service in crash reduction is 270,000 €.

The reduction in traffic volume was approximately 10% when comparing traffic in adverse road weather conditions and in normal conditions one week before or after the adverse conditions (Sihvola and Rämä 2008). However, the reduction does not result entirely from the road weather information service. Anttila et al. (2001) showed that 90% of drivers knew of the road weather information service and 26% of them stated that the service had affected their decision to make the trip. Therefore the reduction in the traffic volume due to the road weather information service can be estimated to be approximately 2%. A 1% change in traffic volume changes the accident rate by 0.5% (Peltola et al. 2005). Therefore a 2% reduction in traffic volume leads to a 1% reduction in traffic accidents. The proportion of adverse road weather was estimated to be 30% of the wintertime. There were approximately 8,270 road traffic accidents per year in total on main roads in 2008, 45% of which in winter time (Peltola et al. 2004), and an average cost of an accident is 118,000 € (Road Administration 2005). Consequently, the benefit of road weather information service is 1,317,400 € in accident costs due to the reduction of traffic volume.

The reduced traffic volume results also in reduced mobility. The cost of the reduced mobility is estimated to be equal to a half the costs of the trips that are cancelled due to the road weather information service. However, in practice all of the 2% reduction in traffic volume is not cancelled trips; some trips are only postponed until road conditions improve. Those trips should not be included in the calculation. Let's assume that 50% of trips are cancelled and 50% postponed. For the cost of reduced mobility driving costs and time costs are considered. The driving costs of passenger cars is 5.68 cnt/km, 7.42 cnt/km of vans, 36.12 cnt/km of trucks and 23.00 cnt/km of buses (Road Administration 2005). Consequently, the cost of reduced mobility can be estimated at approximately 4,100 €.

The valuations and comfort were measured along with willingness to pay and the number of service users and their impact on the economic status of private persons (e.g. changes in service prices) and on corporate economy (Figure 1). Finnish road weather information service has been provided to the public free of charge. The more detailed and personalized road weather information service VARO which was aimed at long distance commercial transport companies

(Damski et al. 2005) did not receive a sufficient number of users willing to pay. Therefore it can be assumed that users are not willing to pay for the road weather information service, either in its current form or in an improved form. Consequently, the service has no impact on the economic status of private persons (e.g. changes in service prices) or on the corporate economy.

The road weather service is well known among drivers as 90% of the drivers in the study carried out by Anttila et al. (2001) knew of it. 62% of drivers interviewed in the study by Sihvola and Rämä (2008) had received/looked for road weather related information before or during the trip. Most important sources of information were radio (42%), television (35%) and internet (14%). Most of the interviews were conducted in adverse road weather conditions. Kilpeläinen and Summala (2002) found out that about 16% of drivers had actively acquired weather information, in addition to which there was a substantial group of drivers who considered that they had received information passively – probably mostly via radio and television. Sihvola and Rämä (2008) studied which kind drivers seek information. The result was that those travelling long distances sought information more often than those travelling short distances, those who drive less than 20 000km/year sought information more often than those driving more than this and young (less than 26 years of age) and old (over 64 years of age) drivers sought for information less than middle aged drivers.

A summary of socio-economic impacts is presented in Table 3. The total cost impact on travel time, predictability, safety, valuations and comfort of the current road weather information service is 13,255,000 €.

Table 3 Most important indicators or weather information services and their impact on costs

Overall objective	**Indicator**	**Socio-economic impact of indicator**	**Impact on costs (€)**
Travel time and predictability	Average speed	3 km/h for service users in adverse road weather conditions	–19,800
	Predictability of travel time	The average speed reduction is compensated by improved predictability of travel time.	19,800
Safety	No. of traffic accidents	100 accidents/year	270,000
	No. of traffic accident injuries	28 injuries/year	6,600,000
	No. of traffic accident fatalities	2.5 fatalities/year	5,071,500
	Veh. km driven	–2%	1,317,400
Mobility	Reduced mobility	–0.5%	–4,100
Valuations and comfort	Willingness to pay	-	–
	No. of service users	active: 16% of drivers active + passive: 62% of drives	–

4.2 Service Quality

Quality of the service was determined based on veracity of the service. The veracity of the service is determined as an error probability that gives a percentage of content provided outside of the stated quality boundaries. Error probability cannot be determined as false events per total number of reported events for the road weather service as, by definition, the road weather class given for an area is the worst road weather on the main road network in a (relatively) large area. Therefore, although the road weather would be normal on all the roads except one with poor conditions, the information given on poor conditions would be correct. Nevertheless, the error probability can be assumed to be correlated with the predictions and observations made on individual sections.

The success of the information has been studied in relation to the number of accidents. The rationale behind this is that information provision can prevent only part of accidents and, therefore, the number of accidents would be larger in adverse weather conditions despite the road weather information provision (Sihvola et al. 2008).

There were 39 days (23%) with elevated accident levels during the winter of 2004–2005 and 8 days (5%) with high accident level (Table 4). A total of 17 days with high or elevated accident level had road weather forecasts classified as normal. 61% of all the accidents happened during days with road weather information given as "normal", 34% "poor" and 6% "hazardous". During the same period the road weather information service forecasted road weather to be normal 66% of time, poor 31% of time and hazardous 3% of time. Days with elevated accident levels were distributed evenly among information classes. However, the days with high accident levels were concentrated on days when road weather information was forecasted as poor or hazardous. (Sihvola et al. 2008)

If the quality of the road weather information service is determined by its ability to warn ahead of those days with high accident levels, the veracity of the service is 99% (i.e. 22% of 8 days wrong in 170 day long period). If the service should also warn of days with elevated accident levels, the veracity is 85% (i.e. 22% of 8 days + 60% of 39 days wrong in 170 day long period). If the latter target is chosen, the error probability of the current system is 15%. However, accident levels have natural variation and do not correspond totally with the veracity of the service. Therefore, expert opinion has to be applied to determine the veracity in terms of error probability of the service.

Table 4 Traffic accidents by road weather class (Sihvola et al. 2008)

Forecast	All accidents	Accidents for days with high accident level	Accidents for days with elevated accident level
Normal	61%	22%	60%
Poor	34%	54%	34%
Hazardous	6%	24%	6%
Number of days	170	8	39

For the first years of the road weather information system (status “1998”) the error probability of road weather information (Table 5) was estimated to be approximately 14% (Pilli-Sihvola 2009). Most of the preceding studies have been based on this status.

In 2003, the system was improved by the investment in updated road weather stations (from Vaisala MILOS to ROSA) with partly improved sensors and improved analysis software. Only the studies by Aittoniemi (2007) and Hautala and Leviäkangas (2007) can be regarded as referring to the status of road weather related information services around the "2003" situation, having the error probability of approximately 11% (Pilli-Sihvola 2009).

Table 5 Implementation of service at various levels of data quality

Quality level	Error probability	Description of quality
1998	14%	Basic system
2003	11%	System improved by the investment in updated road weather stations (from Vaisala MILOS to ROSA) with partly improved sensors and improved analysis software

4.3 Costs

The main costs of the service quality improvement from the "1998" situation to "2003" situation was the investment in updated road weather stations (from Vaisala MILOS to ROSA) with partly improved sensors and improved analysis software. According to Kulmala and Karhumäki (2007), the investment costs of the update were approximately 18,000 € per station, whereas the costs of a new station were 1.5 times that of the update. The annual maintenance and operation costs of a road weather station were approximately 2,000 €. Note that the updates have been carried out mainly due to the requirements of winter maintenance.

According to an unpublished Finnish survey of traffic management related costs from 2001, the annual costs of the road weather system were around 2.5 million euro, of which approximately 1.3 M€ were due to the data collection related costs including reinvestment costs. Note that road weather stations need to be replaced in any case in 8–12 years. If we assume that the new improved stations increased the data collection costs annually by around 50%, the new road weather data collection costs of the "2003" situation were about 2 M€ annually.

If we assume that the share of the road weather system costs to be allocated to road weather related traffic information would be around 10% of the total, this would mean in the two cases 0.25 M€ and 0.32 M€ annually. The education of traffic management personnel (50 persons) and contractors (50 persons) related to changes in road weather system has been approximately 0.15 M€ in total (1,500 €/person). The annual costs of road weather information service provision would be in both cases about 1.5 M€ annually. The costs of the alternatives are summed up in Table 6. The road weather model and data fusion algorithms used have also

Table 6 The costs of the alternatives

Costs	Quality level	
	1998	**2003**
Annual road weather system costs (M€)	2.5	3.2
Weather information share of road weather system costs (M€)	0.25	0.32
Education of TMC personnel + contractors related to changes in road weather system (M€)		0.15
Annual weather information service provision costs (M€)	1.5	1.5
Total annual costs for road weather information service (M€)	1.75	1.97

been updated between 1998 and 2003. However, no detailed information was available on the costs related to improving road weather model nor updating the 'data fusion algorithms' nor the share of these costs to be allocated to the road weather information service.

4.4 Quality vs. Benefits

Concerning the quality of service expressed as error probability, most of the Finnish studies have been carried out based on the status of "1998", at the beginning of the Finnish road weather service. At that time the error probability of road weather information was estimated to be around 14%. On the basis of the behavioral and safety impact studies, the likely reduction of crashes involving injury in poor and hazardous road weather conditions could be around 8% on the main road network.

Only the studies of Aittoniemi (2007) and Hautala and Leviäkangas (2007) can be regarded as referring to the status of road weather related information services around the "2003" situation, having the error probability of around 11%. On the basis of these studies, the likely reduction of crashes involving injury in poor and hazardous road weather conditions could be around 10% on the main road network.

Hautala and Leviäkangas (2007) estimate that around half of the safety impact of weather information is due to speed reduction. Nilsson (2004) showed that the speed reduction's safety effect for fatal accidents is equal to the squared safety effect for crashes involving injury. On the basis of these studies, we can estimate the effects on fatal crashes in poor and hazardous conditions to be 12% for the 1998 situation and 15% for the 2003 situation.

The annual number of crashes involving injury in poor and hazardous conditions on main roads in 2008 was around 200 and the corresponding number of fatal crashes was around 15. Hence, the annual benefits of road weather information of the two error probabilities were estimated to be 6.6 M€ for error probability 14% and 11.7 M€ for error probability 11% (Table 7), using the Finnish accident involving injury unit cost of 330,000 € and 2,205,000 € for fatal accidents.

Table 7 Service impacts at different quality levels

Quality level	Error probability	Reduction of injury accidents (%)	Annual reduction of injury accidents (number)	Reduction of fatal accidents (%)	Annual reduction of fatal accidents (number)	Annual saving in injury accident costs (M€)
1998	14%	8	16	12	1.8	6.6
2003	11%	10	20	15	2.3	11.7

Cost-benefit ratio was calculated at the two levels of data quality (Table 8). Discount rate of 5% and lifetime of 8 years for service were assumed in the cost-benefit analysis according to Viking Evaluation Guidelines. Quality level "2003" had higher benefit cost ratio (6.4) than the quality level "1998" (3.8).

Table 8 Cost-benefit ratio at various levels of data quality

Quality level	Benefits [NPV/€]	Costs [NPV/€]	B/C ratio
1998	44,790,000 €	11,876,000 €	3.8
2003	79,401,000 €	12,351,000 €	6.4

5 Discussion

The objective is to study the optimum level of service quality for the Finnish road weather information service and to test the methodology to find the optimum service quality developed in the QUANTIS project.

The limited availability and variation in the quality of evaluation studies on traffic information services seems to be a barrier to analyzing the dependence of service benefits and costs on service quality. Because commonly accepted indicators for data and service quality have not been available until recently, the level of data or service quality has not been documented in most evaluation studies focused on traffic information services. This means that there are only a few studies in which both service impacts and service quality have been documented which could be used to make generalized conclusions about the relationship between service quality and service impacts and costs. Given the limited availability of suitable evaluation studies, information about the impacts and costs of a service at various levels of service quality usually have to be collected through expert interviews.

The study showed that it is not always easy to find meaningful interpretation for quality attributes defined in standards related to data or service quality. Because all services do not have formal definitions for all aspects of service quality, the interpretations for the quality attributes in a particular case may have to be defined during the evaluation process.

In case of the Finnish road weather information service, the relation between service quality costs turned out to be complex. Because the service output is produced on the basis of several data sources in a process involving intelligent data fusion performed by computers and human users, the results of changes in the service production process are not always easy to predict. For example, human users may compensate for the unavailability of one data source by using his or her experience with similar situations and information available from the remaining data sources.

The approach presented above can provide estimates for the impacts and costs of service in service cases in which there is information available about the service production process, dissemination methods and users. However, this does not guarantee that the results are transferable from one service case to another. For example, the costs to implement a service may be significantly different in different operating environments.

Cost-benefit analysis of a service provides the monetary values of benefits and costs and the benefit-cost ratio at various levels of service quality. However, the interpretation of the results of the cost-benefit analysis depends on the goals of the decision-maker. For example, some decision-makers may want to implement projects with benefit-cost ratio of less than one if the likely outcome of the project is in line with their political objectives. For example, the "vision zero" adopted in the Northern Countries for traffic safety may lead experts to recommend measures which most probably improve safety but whose socio-economical profitability in terms of benefit-cost ratio cannot be guaranteed (Rajamäki et al. 2008).

6 Conclusions

In the case of the Finnish road weather information service, the quality level "2003" with an error percentage of 11% has a higher cost-benefit ratio than the quality level "1998" with an error percentage of 14%. In both analyzed cases, most socio-economical benefits come from improved safety of road users which is the main objective of the service under analysis. For these reasons, the quality level "2003" should be preferred over the earlier one.

However, there is some doubt as to whether the increase in service impacts has been caused solely by the improvement in the error probability of the service. On one hand, it is probable that the most changes in impacts are a result of the improved error probability of the service output, since this has been the most significant change in service quality between years 1998 and 2003. On the other hand, changes in the levels of other quality attributes such as service timeliness have not been analyzed in detail.

Given the differences between ITS services and decision-making situations, it is not likely that any simple criteria which would be applicable in all situations can be given to select the optimum service quality.

References

1. Hautala, R., Leviäkangas, P. (eds.): Ilmatieteen laitoksen palveluiden vaikuttavuus. Hyötyjen arviointi ja arvottaminen eri hyödyntäjätoimialoilla [Effectiveness of Finnish Meteorological Institute (FMI) services]. Espoo 2007. VTT Publications 665.205 + p. 73 http://www.vtt.fi/inf/pdf/publications/2007/P665.pdf (accessed May 21, 2010)
2. Öörni, R., Kulmala, R., Newton, D., Scheider, T., Kellermann A.: Deliverable 1 of QUANTIS - Definition of key European ITS Services and Data Types, Vienna, Austria (2009), http://www.quantis-project.eu/2009-11-19_QUANTIS_D1_final.pdf (accessed May 17, 2010)
3. Scheider, T., Newton, D., Öörni, R., Kellermann, A.: Report on service benefits and costs. Deliverable D2 of QUANTIS, version 0.9, unpublished draft version (2010)
4. Aittoniemi, E.: Tieliikenteen tietopalveluiden vaikutukset henkilövahinko-onnettomuuksiin (2007) [Effects of road traffic information services on injury accidents.] Master's Thesis Draft (February 15, 2007)
5. Anonymous, Data quality in ITS systems. ISO Technical report ISO/TR 21707:2008, International Organization for Standardization (2008)
6. Anonymous: QUANTIS – Quality assessment and assurance methodology for traffic data and information services web site (2010), http://www.quantis-project.eu (accessed May 17, 2010)
7. Anttila, V., Nygård, M., Rämä, P.: Liikennesää-tiedotuksen toteutuminen ja arviointi talvikaudella 1999-2000 (2001) [Evaluation of the Road Weather Information Service in winter 1999-2000], Tiehallinnon selvityksiä 41/2001, Road Administration, Helsinki, p. 77
8. Damski, J., Hippi, M., Horppila, H., Innamaa, S., Kangas, M., Kiuru, J., Lamminluoto, M., Nurmi, P., Plathan, P., Saukkonen, L., Strahlendorff, M.: Kuljettajien varoituspalvelu. Palvelun sisältö ja tekniset määrittelyt AINO-julkaisuja : 6/2005, Liikenne- ja viestintäministeriö, Helsinki, p. 69 (2005)
9. Kangas, J., Kärki, J.-L.: Autojen nopeudet pääteillä sekä yhdysteillä vuonna 2008. Tiehallinnon selvityksiä 15/2009, Finnish Road Administration, Helsinki, p. 70, 54 (2009)
10. Kilpeläinen, M., Summala, H.: Kelitiedotuksen kokeminen ja vaikutukset. [Road user's perceptions and the impacts of the road weather information service in Finland.] Helsinki 2002. Finnish Road Administration. Finnra Reports 59/2002. 42, p. 4 (2002)
11. Kulmala, R., Karhumäki, T.: VIKING Monitoring Plan 2006. Helsinki, VIKING. 55 s. + liitt. 4 s VIKING MIP2006 Deliverable, Version 1.01 (2007)
12. Nilsson, G.: Traffic Safety Dimensions and the Power Model to describe the effect of speed and safety. Bulletin 221. Department of Technology and Society. Lund University. Sweden (2004)
13. Peltola, H., Rajamäki, R., Malmivuo, M.: Talviajan nopeudet ja raskas liikenne. Liikenne- ja viestintäministeriön julkaisuja 67/2004, liikenne- ja viestintäministeriö, Helsinki, p. 86 (2004)
14. Peltola, H., Rajamäki, R., Rämä, P., Luoma, J., Beilinson, L.: Tieliikenteen turvallisuustoimenpiteiden arviointi ja kokemukset turvallisuussuunnitelman laatimisesta. LINTU-julkaisuja 1/2005, Ministry of Transport and Communications, Helsinki, p. 142 (2005)

15. Pilli-Sihvola, Y.: An interview with road weather expert Yrjö Pilli-Sihvola, Finnish Road Administration, Kouvola (2009)
16. Rajamäki, R., Innamaa, S., Peltola, H.: Esiselvitys katsastustoiminnan vaikutuksista liikenneturvallisuuteen. Ajoneuvohallintokeskuksen tutkimuksia ja selvityksiä 6/2007 (2008), http://www.ake.fi/NR/rdonlyres/99444215-9D6D-4E1F-BA4F-6308F0CC50D/0/AKE607Katsastuksenvaikutusliikenneturvallisuuteen.pdf (accessed May 17, 2010)
17. Road Administration: Tieliikenteen ajokustannusten yksikköarvot 2005. Road Administration, Helsinki, p. 15 (2005)
18. Road Administration: Tieliikenteen suoritteet vuonna 2008, Tiedote 20.2.2009, Finnish Road Administration, Helsinki (2009)
19. Rämä, P., Kulmala, R.: Effects of variable message signs for slippery road conditions on driving speeds and headways. Transportation Research Part F 3, 85–94 (2000)
20. Rämä, P., Raitio, J., Harjula, V., Schirokoff, A.: Sää- ja kelitietoon perustuvan liikenteenohjausjärjestelmän vaikutukset yksiajorataisella osuudella valtatiellä 7. Helsinki, Tiehallinto. 68 s. + liitt.11 s. Tiehallinnon selvityksiä 44 (1999) ISBN 951-726-588-3
21. Rämä, P., Raitio, J., Anttila, V., Schirokoff, A.: Effects of weather controlled speed limits on driver behaviour on a two-lane road. In: Proceedings of Traffic safety on three continents, International Conference, Moscow, September 19-21, TRB, CSIR, VTI (2001)
22. Rämä, P., Schirokoff, A., Rajamäki, R.: Muuttuvien nopeusrajoitusjärjestelmien turvallisuus [Safety of Variable Speed Limit Systems]. Tiehallinto. 46 S. + Liitt. 8 S. Tiehallinnon Selvityksiä 54 (2003) http://alk.tiehallinto.fi/Julkaisut/pdf/3200841_Vms-Turva.Pdf.Helsinki Isbn 951-803-154-1
23. Sihvola, N., Rämä, P.: Kuljettajien käsityksiä kelistä ja kelitiedotuksesta – tienvarsihaastattelu talvikelissä [Driver's opinions about road weather conditions and road weather information – interviews during winter conditions]. Tiehallinnon selvityksiä 16/2008, Finnish road administration, Helsinki, p. 10, 62 (2008)
24. Sihvola, N., Rämä, P., Juga, I.: Liikennesäätiedotuksen toteutuminen ja arviointi 2004–2007 ja yhteenveto 1997–2007. [Determining the successfulness of the road weather information service 1997–2007]. Tiehallinnon selvityksiä 15/2008, Finnish Road Administration, Helsinki, p. 18, 84 (2008)

Using Vehicles as Mobile Weather Platforms

Sheldon Drobot, Michael Chapman,
Paul A. Pisano, and Benjamin B. McKeever

Abstract. One of the goals of the Research and Innovative Technology Administration's IntelliDrive^{SM} initiative is for the public and private organizations that collect, process, and generate weather products to utilize vehicle sensor data to improve weather and road condition products. It is likely that some users will not be able to contend with the complexities associated with vehicle data, such as data quality, representativeness, and format. A solution for addressing this issue is to utilize a Vehicle Data Translator (VDT) to pre-process weather-related vehicle data before it is distributed to data subscribers. This paper will describe the VDT and how vehicle data sets are processed by the prototype VDT to generate derived weather and road condition information.

1 Introduction

The utilization of data from mobile platforms is not new in the weather community, but the utilization of data from vehicles poses significant technical

Sheldon Drobot
National Center for Atmospheric Research,
3450 Mitchell Lane, Boulder, CO 80301, 303-497-2705
e-mail: drobot@ucar.edu

Michael Chapman
National Center for Atmospheric Research,
3450 Mitchell Lane, Boulder, CO 80301, 303-497-8305,
e-mail: mchapman@ucar.edu

Paul A. Pisano
USDOT/Federal Highway Administration, 1200 New Jersey Avenue,
SE, Washington, D.C. 20590, 202-366-1301
e-mail: paul.pisano@dot.gov

Benjamin B. McKeever
USDOT/Research and Innovative Technology Administration,
1200 New Jersey Avenue, SE, Washington, DC, 20590, 202-366-4876
e-mail: ben.mckeever@dot.gov

J. Düh et al. (Eds.): Data and Mobility, AISC 81, pp. 203–213, 2010.
springerlink.com © Springer-Verlag Berlin Heidelberg 2010

challenges (1–3), particularly with respect to data quality. Nevertheless, vehicle-based probe data from initiatives such as IntelliDrive^{SM} will significantly increase the density of weather observations near the surface and will provide unique datasets for deriving and inferring road and atmospheric condition information, once issues such as deployment of roadside equipment and vehicle connectivity are addressed.

The amount of data that would potentially flow through a vehicle-based data network such as IntelliDrive^{SM} could be immense. It is likely that many prospective users will not be capable of handling the vast quantities of data that will be generated. Applications must be implemented to facilitate the use of vehicle-based weather data, because without such a function, the feasibility of utilizing vehicle probe data will be lower and there will be a substantially greater risk in its use.

One possible solution for mitigating the adverse impacts of weather on the transportation system is to provide improved road and atmospheric hazard products to road maintenance operators and the travelling public. With funding and support from the U.S. Department of Transportation's (USDOT) Research and Innovative Technology Administration (RITA) IntelliDrive initiative and direction from the Federal Highway Administration's (FHWA) Road Weather Management Program, the National Center for Atmospheric Research (NCAR) is conducting research to develop a Vehicle Data Translator (VDT) that incorporates vehicle-based measurements of the road and surrounding atmosphere with other, more traditional weather data sources, and creating road and atmospheric hazard products for a variety of users.

This paper outlines recent progress on the development of a VDT.

2 Vehicle Probe Data - Weather Data Processing

Vehicle data can be complex and can pose a significant analytical challenge, particularly when it comes to measuring or deriving weather and road condition data. For example, some obvious issues center on how to deal with the large data volume, the timeliness of the data, data quality and representativeness, and data format(s). These issues are not dissimilar to those associated with other fixed meteorological datasets, but the complexities are compounded because end users will have little knowledge about the data source. In contrast, the National Weather Service (NWS), Federal Aviation Administration (FAA), and other traditional providers of weather data adhere to stringent standards for instrumentation accuracy, precision, and siting.

The uncertainties and complexity associated with raw vehicle data (i.e. unprocessed data from vehicles) is likely to deter many end users from using the data. As noted above, it is likely that most end users will not be able to handle the immense volume of vehicle data, let alone deal with data quality questions. We anticipate that many, if not the majority, of users of vehicle weather data will require *processed* data. In this context, *processed* data means vehicle data that is extracted from the network of vehicle probe data, quality checked, and disseminated to active data subscribers in near real time. It is also likely that, due to the volume of data, many users will prefer statistically-derived data representing specific geographical areas (such as road segments) or times.

3 The NCAR Vehicle Data Translator Concept

In a fully functional IntelliDrive^{SM} environment, millions of vehicles will be acting as probes and will continuously send reports to the vehicle data network. The concept of the weather and road condition data processor for vehicle data has been discussed in both *Clarus* and Vehicle Infrastructure Integration (VII) initiative meetings (VII was the predecessor to IntelliDrive^{SM}). In our view, not only is the concept sound, but the need for such a function is critical.

A conceptual illustration of the primary processing components of the proposed VDT is shown in Figure 1. Data from IntelliDrive^{SM}-equipped vehicles (e.g., personal and fleet vehicles) is communicated to the Road Side Units (RSEs) when the vehicles are within range of the receivers. The RSEs are connected to the IntelliDrive^{SM} communications network where most of the data will flow. Individual processing components of the VDT are described in the following sections.

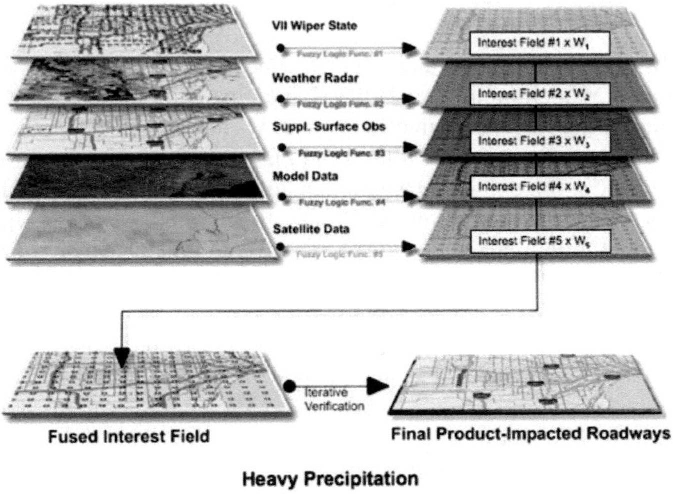

Fig. 1 Vehicle Data Translator (VDT) Conceptual Illustration

3.1 Data Parser

The proposed VDT will include a data parser function that will extract relevant weather and road condition vehicle probe fields from the vehicle data network. The data elements selected for extraction will be determined based on research results and feedback from stakeholders in both the atmospheric and surface transportation communities. The data elements could be added or subtracted as needs vary. The data flowing out of the data parser is still considered raw because it has not been processed in any way.

3.2 Data Filtering Algorithms

Data filtering algorithms will then be applied to the chosen data elements to remove data that is not likely to be representative of the true conditions. For example, outside air temperature measurements may not be representative of the true ambient conditions if the vehicle speed is less than 25 mph (4). Therefore, one test that could be applied to the vehicle data is to discard all outside air temperature data measurements when the vehicle speed is less than 25 mph.

Filters could also be applied to data collected at particular locations that are known to generate errors (e.g., data measurements from inside tunnels). The process of deciding when and how to filter data will require considerable research and will need to be carried out with great care, as one would not want to remove data that may have some value.

3.3 Data Quality Checking Algorithms

One benefit of removing data that is considered unrepresentative in the filtering procedure is a reduction in the data volume that will need to undergo more complex and computationally intensive quality checking (QC) procedures. The QC tests will include many of the common tests that are applied to surface weather data and more complex tests to handle road condition data. In both cases, ancillary data, such as surface weather observations, satellite, radar, climatological data, and model output statistics will be required to conduct many of the quality tests. However, the VDT QC process will need to be sufficiently quick to ensure minimal latency.

The quality checking methods used in the VDT are related to some of those described in the *Clarus* system (5), and include the following:

- Sensor Range Test (SRT)
- Climatological Range Test (CRT)
- Neighboring Vehicle Test (NVT)
- Neighboring Surface Station Test (NST)
- Model Analysis Test (MAT)
- Remote Observation Test (ROT)

The SRT will be employed to identify observations that fall outside the range of the known sensor hardware specifications. Unlike the *Clarus* system, data processing algorithms included in the VDT will not have the advantage of "knowing" what type of sensor (e.g., make, model) produced each measurement. In order to develop an effective SRT, it will most likely be necessary to conduct research on the automotive sensors that are available in the market place and make an educated assessment of the reasonable bounds for this test.

The CRT test will identify observations that fall outside of location-specific climatological (or historical) ranges. This is a more complex task than the SRT because of the variability of the climatological range values over various times, dates, locations, and seasons. Once a climatological range has been assigned for a specific road segment or geographic region (e.g., based on weekly values), the test will simply allow for observations within that range and flag up observations that fall outside of the range.

The NVT is a "nearest neighbor" test which will compare the observation to neighboring vehicles in the road segment. If the observation value falls outside of a dynamic threshold, then the observation will fail the test. The threshold will probably be determined by the number of neighboring observations that are available for comparison. In other words, the higher the quantity of vehicles in a road segment, the tighter the threshold range.

The NST will compare neighboring surface station observations (e.g., Road Weather Information System (RWIS), Automated Surface Observing Systems (ASOS), Automated Weather Observing Systems (AWOS)) to each vehicle observation. This test will obviously depend on the availability of a surface observation within a suitable distance of the road segment.

The MAT will compare observations with a model surface analysis, such as the Real-Time Mesoscale Analysis (RTMA) or the Rapid Update Cycle (RUC) Surface Assimilation Systems (RSAS). This test will compare the observed value to either a range of grid values along a road segment or within a predetermined threshold of the grid value closest to the location of the vehicle.

The ROT will compare vehicle observations to remotely sensed data from either satellite and/or radar. Both the radar and satellite can be used as either a complimentary step with the other QC tests or as the final option if no neighboring vehicles or surface stations are available. As an example, pavement temperature, rain and sun sensor observations are good candidates for this test.

In the QC process, data quality flags will be applied to the raw vehicle data so that data subscribers will have the flexibility of utilizing the raw data or taking advantage of the quality checking flags. After the QC procedure, some data will flow directly to the output queue to minimize data latency (Figure 1).

3.4 Statistical Processing and Derived Variables

Some of the data will be cached and processed to generate statistical values for a given location (e.g., grid cell or point) and time period. The statistical processing will create two separate data streams: "processed data" and "derived data". To generate processed data, the application of a statistical technique (e.g. mean, median) to a set of observations over a known grid segment is performed, and should result in a more robust sample and reduce the overall data load for users that either cannot handle, or do not require individual vehicle data. To generate derived data, a more computationally intensive procedure is necessary. This procedure focuses on deriving new or enhanced road and atmospheric variables of interest to the surface transportation community (6). The derived variables will be comprised of a blend of weather and non-weather related IntelliDrive^{SM} data elements in conjunction with ancillary data sets. Table 1 lists possible derived observations that are currently being considered during the incipient phases of IntelliDrive^{SM}, the ancillary data required to construct each derived observation, and the relevant vehicle observations (observed and input).

An evaluation of the optimum road segment length will be performed to determine the appropriate value that will be required to generate a spatially representative derived observation value. After the road segment size is determined, a step will be needed to evaluate whether the number of observations over the valid road segment is sufficient to derive an observation. For example, if a road segment value is produced by incorporating data from only one vehicle at a given time, it is safe to assume that the confidence in that value will be significantly lower than if the value for a given segment and time was derived from multiple observations from multiple sources. It will be important to identify the ideal minimum number of

Table 1 Sample Matrix for VDT Derived Weather-Related Observations

Derived Observation	Ancillary Data	Vehicle Observation
Ambient Air Temperature	RSAS, Surface Stations	Vehicle Speed, Temperature, Hours of Operation
Barometric Pressure	RSAS, Surface Stations	Vehicle Speed, Barometric pressure, Hours of Operation
Precipitation Occurrence	Radar, Surface Stations, Satellite	Wiper Status, Vehicle Speed, ABS, Stability Control, Traction Control, Headlight Status
Precipitation Type	Radar, Surface Stations, RSAS, Sounding	Wiper Status, Vehicle Speed, Temperature, Elevation, ABS, Stability Control, Traction Control
Precipitation Rate	Radar, Surface Stations	Wiper Status, Rain Sensor, Vehicle Speed, ABS, Stability Control, Traction Control, Headlight Status
Fog	Radar, Surface Station Visibility, Satellite Cloud Classification	Wiper Status, Rain Sensor, Vehicle Speed
Pavement Conditions (wet, dry, icy)	Radar, Surface Stations	Vehicle Speed, ABS, Stability Control, Traction Control, Temperature, Wiper Status
Boundary Layer Water Vapor	Radar, Soundings, Surface Stations	Temperature, Wiper Status
Pavement Temperature	Surface Stations	Temperature, Vehicle Speed
Smoke	Satellite Smoke Algorithm, Surface Station Visibility,	Vehicle Speed, Headlight Status

observations during the early stages of VDT development; this number will differ for each type of observation.

It should also be noted that the merging of different data sources (vehicle observations and ancillary data) will require the use of expert systems (e.g., fuzzy logic, neural networks) and/or decision tree algorithms in order to produce robust observations. It is likely that a combination of techniques will be required for some of the products.

4 VDT Display

A key component of the VDT concept includes the capability to display the data. A prototype display is under development to support VDT research and to convey the utility of vehicle-based data (Figure 2). The VDT display is browser-based and is capable of rendering the following data types:

- Vehicle probe messages
- Standard surface weather observations
- Gridded radar data
- RSAS data
- Satellite cloud mask data
- Road segment statistics (data range, variance, means, modes, etc.)
- Derived road segment values
- National Weather Service watch and warning data
- Twitter #wxreport messages

5 VDT Product Examples

This section outlines two examples of VDT products. The first is a conceptual description of a derived fog product, and the second is a conceptual description of a precipitation product.

5.1 Fog Example

Fog remains one of the more difficult-to-diagnose road weather hazards. A schematic of a decision tree for the derivation of fog is presented in Figure 3. Probe data will provide useful observations of the status of the vehicle (e.g. wipers, vehicle speed) and observations of the atmosphere (e.g. temperature, rain sensor). These observations can be combined with ancillary data (e.g. radar, visibility and humidity from surface stations, and satellite cloud classification information) to diagnose the likelihood of fog. For example, if a vehicle's (and surrounding vehicles') speed decreases rapidly relative to the posted speed limit of a road segment, a decision tree can be used to determine if the decrease in speed was caused by fog, precipitation, or some other non-weather-related factor. The

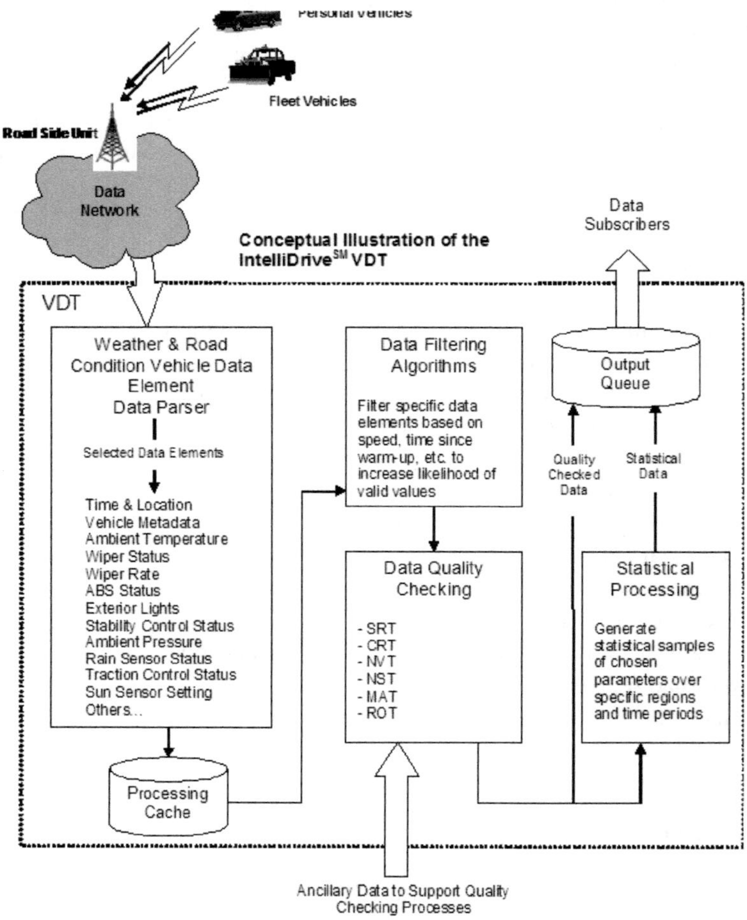

Fig. 2 Example of prototype VDT display

first test would determine the ratio of the rate of speed versus the speed limit of the road segment. If this ratio equals a certain threshold (e.g. 50%) then the algorithm would next consider the headlight status observation. If the headlights were on and the date and time suggest that it is daytime, then the algorithm would consider the wiper status. The algorithm would next check the relative humidity and visibility from the closest (in time) surface station observation report. If visibility is less than a predetermined threshold (e.g. ¼ mile) and humidity is high then there is some confidence that fog exists over the road segment. If a surface station is not available or if the observation from the surface station is outside of a predetermined temporal threshold then the algorithm could employ radar and satellite data. In this event, if the radar shows no precipitation but the cloud classification algorithm diagnoses a low-level stratus cloud, a fog classification could occur. While the above description assumes a decision tree for the

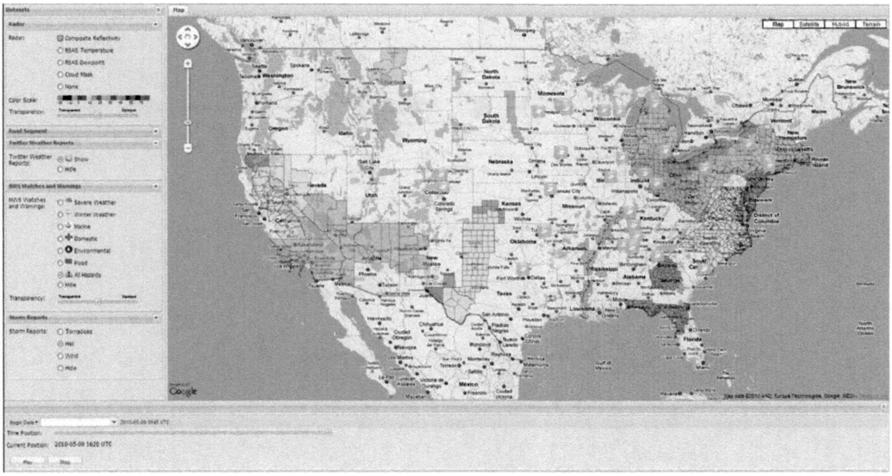

Fig. 3 Schematic of decision tree algorithm for dense fog diagnostic application

combination of these data sources, a more robust fuzzy logic technique could also be employed when it is determined that applying discreet thresholds on the observations is not ideal.

5.2 *Precipitation Example*

Another major hazard for the surface transportation industry is precipitation, due to the negative effects of precipitation on the tire/roadway interface (i.e. lower friction, hydroplaning). Diagnosing the occurrence, rate, and type of precipitation is a necessity if improvements to the safety and efficiency of the roadway are to be realized in the future. IntelliDrive^{SM}-enabled weather observations have the potential to provide the high density, real-time observations that could fill in the gaps between radar and the less dense fixed surface weather station network, and also add valuable information concerning the effects that the precipitation is having on the vehicle itself (e.g. ABS, activation of Traction Control, reduction in speed, collisions, etc.).

Figure 4 is a schematic, which represents the design for a precipitation algorithm using fuzzy logic. The algorithm would combine Wiper State with radar, surface observations, model data (e.g. RSAS), and satellite at different interest levels. Each variable would then be given a weight and fused into a consolidated interest field for precipitation. The final product would be a mapping of impacted road segments.

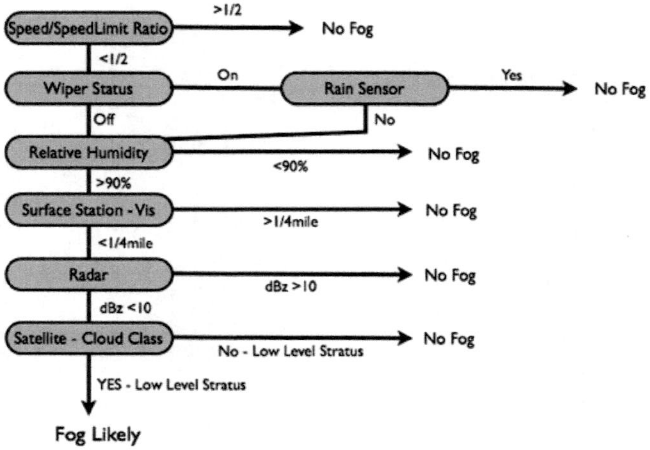

Fig. 4 Schematic of the combination of vehicle data with ancillary data for producing a precipitation occurrence algorithm

6 Conclusions

From a weather perspective, the overarching goal of the IntelliDrive^{SM} initiative is for the public and private organizations that collect, process, and generate weather and transportation products to utilize vehicle data to improve weather and road condition products, and then to provide those products to transportation system decision-makers, including travelers. Nonetheless, the utilization of data from vehicles poses significant technical challenges, particularly with respect to data quality and quantity. The amount of data potentially flowing through a vehicle-based data network, such as IntelliDrive^{SM}, could be immense and it is likely that many prospective users will not be capable of handling the vast quantities of data that are expected.

The VDT discussed in this paper is one approach to pre-processing weather-related vehicle data before it is distributed to data subscribers. The function of the VDT is to extract the data elements needed to derive weather and road condition information from vehicle probe data, filter the data to remove samples that are likely to be unrepresentative, quality check the data utilizing other local surface observations and ancillary datasets, generate statistical output for specific road segments, and disseminate the quality-checked and statistically processed data to subscribers, which may include other data processing and dissemination systems such as the U.S. Department of Transportation's (USDOT) *Clarus* System.

Acknowledgements. Research conducted during this project was funded by FHWA Contract DTFH61-08-C-00004.

References

1. Srinivasan, K.K., Jovanis, P.P.: Determination of Number of Probe Vehicles Required for Reliable Travel Time Measurement in Urban Network. Transportation Research Record: Journal of the Transportation Research Board, No. 1537, Transportation Research Board of the National Academies, Washington, D.C., 15–22 (1996)
2. Cheu, R.L., Qi, H., Lee, D.-H.: Mobile Sensor and Sample-Based Algorithm for Freeway Incident Detection. Transportation Research Record: Journal of the Transportation Research Board, No. 1811, Transportation Research Board of the National Academies, Washington, D.C., 12–20 (2002)
3. Shladover, S.E., Kuhn, T.M.: Traffic Probe Data Processing for Full-Scale Deployment of Vehicle–Infrastructure Integration. Transportation Research Record: Journal of the Transportation Research Board, No. 2086, Transportation Research Board of the National Academies, Washington, D.C., 115–123 (2008) DOI: 10.3141/2086-14
4. Stern, A.D., Pisano, P., Kennedy, P., Petty, K., Mahoney, W.: A next generation of observations based on passenger vehicles. published in the 88th Annual Meeting of the American Meteorological Society Proceedings American Meteorological Society Annual Meeting, New Orleans, Louisiana, January 24 (2008)
5. United States Department of Transportation, Clarus Weather System Design: System Design Description Version 1.1, USDOT, ITS Joint Program Office. Washington, D.C (2006)
6. Petty, K.R., Mahoney, W.P.: Weather Applications and Products Enabled Through Vehicle Infrastructure Integration (VII): Feasibility and Concept Development Study. In: National Center for Atmospheric Research, Boulder, Colorado, p. 116 (2007)
7. McGukin, T.: Small town Attitude. ITS International NA Edition, pp. NA5–NA6 (November/December 2007)

Appendix

Host and Organizer of the Lakeside Conference

The 2010 Lakeside Conference Data and Mobility, Transforming Information into Intelligent Traffic and Transportation Services, is a co-operative venture of the ATTC - Austrian Traffic Telematics Cluster, AustriaTech - the Federal Agency for Technological Measures Ltd., the U.S. Commercial Service Austria and the Lakeside Science & Technology Park.

Lakeside Science & Technology Park (Host)

The Lakeside Science & Technology Park is a platform for co-operation between companies and university institutions in the field of information and communication technologies (ICT). It offers a place of interdisciplinary research and development, training, production and services. Every year, the Park management organizes an international conference, whose theme alternates between a special IT topic (Safety in Mobility, Intelligent Weather Information Systems and Services in Traffic and Transport, 2008) and a business topic (Knowledge Loves Company, Co-operation Strategies between Universities and Companies, 2009). The special IT topics have been organized jointly with the above partners.

AustriaTech - Federal Agency for Technological Measures Ltd.

AustriaTech - Federal Agency for Technological Measures Ltd. was set up in 2005 by the Austrian Ministry of Transport, Innovation and Technology (BMVIT). AustriaTech is opening up the optimum, non-profit-making benefits of telematics in the transport system and its aim is to stimulate the development of ITS ("Intelligent Transport Systems") in new areas of activity.

Specific concerns are the development and deployment of suitable technologies to ensure efficient transport for the future, and - working alongside the infrastructure operators - the organisation of transport on an intermodal basis.

ATTC - Austrian Traffic Telematics Cluster

The Austrian Traffic Telematics Cluster (ATTC), which consists of 16 renowned Austrian companies, has set the goal of working both on the further development as well as the practice-oriented implementation of new technologies in the area of telematics systems for traffic management.

U.S. Commercial Service Austria

The U.S. Commercial Service Austria at the American Embassy offers assistance to exporters and importers of U.S. goods and services to Austria. Its trade specialists can help you identify trade opportunities, find trading and technology partners, launch your company, obtain market research, and work with you to achieve your business goals.

Patronage

Hans Schönegger - Chief Executive Officer Lakeside Science & Technology Park
Reinhard Pfliegl - Chief Executive Officer AustriaTech
Alois Schedl - ATTC President
Christopher Quinlivan - Counselor for Commercial Affairs, American Embassy in Austria

Conference Organization Team

Conference Chair
  Maria Mack - Lakeside Science & Technology Park
Organization Chair
  Roman Ulram - Lakeside Science & Technology Park
Call for Papers and Programme Chair
  Julia Düh - AustriaTech
Conference Committee
  Marta Haustein - U.S. Commercial Service, American Embassy in Austria
  Helmut-Klaus Schimany - ATTC
  Hartwig Hufnagl - ATTC

Scientific Committee

Julia Düh - AustriaTech, Austria (Chair)
Christian Faisstnauer - Autostrada del Brennero, Italy
Martin Fellendorf - University of Technology Graz, Austria
Susanne Fuchs - Brimatech Services, Austria
Heinrich Garn - AIT Austrian Institute of Technology, Austria
Antonio Marques - ETRA, Spain
Peter Maurer - AIT Austrian Institute of Technology, Austria
Adam Nagy - State Motorway Management, Hungary
Roberto Edoardo Nenz - IMM, Italy
Reinhard Pfliegl - AustriaTech, Austria
Alexander Sauter - Austro Control, Austria
Emil Simeonov - FH Technikum Wien, Austria
Klaus Stadlbacher - Central Institute for Meteorology and Geodynamics, Austria
Gwenaëlle Toulminet - ASFA/Ecole des Mines, France

Author Index

GPSR Compliance

The European Union's (EU) General Product Safety Regulation (GPSR) is a set of rules that requires consumer products to be safe and our obligations to ensure this.

If you have any concerns about our products, you can contact us on ProductSafety@springernature.com

In case Publisher is established outside the EU, the EU authorized representative is:

Springer Nature Customer Service Center GmbH
Europaplatz 3
69115 Heidelberg, Germany

Batch number: 10755023

Printed by Printforce, the Netherlands